T0292510

The Natural
HISTORY
of the Media Luna
Watershed

Richard Worfel

assisted by
Ossiel Martinez

Illustrated by
Israel Ivan Garcia Martinez

Copyright © 2015 by Richard Worfel.   741332
Library of Congress Control Number: 2016909865

ISBN: Softcover   978-1-5245-1024-4
        Hardcover   978-1-5245-1026-8
        EBook   978-1-5245-1023-7

All rights reserved. No part of this book may be
reproduced or transmitted in any form or by any means,
electronic or mechanical, including photocopying,
recording, or by any information storage and retrieval
system, without permission in writing from the copyright
owner.

Print information available on the last page

Rev. date: 07/23/2016

To order additional copies of this book, contact:
Xlibris
1-888-795-4274
www.Xlibris.com
Orders@Xlibris.com

Table of Contents

FORWARD IX

PREFACE XI

Chapter 1   INTRODUCTION 1

Chapter 2   DESCRIPTION OF THE MEDIA LUNA SYSTEM 4

Chapter 3   THE CHANGING ENVIRONMENT OF MEDIA LUNA 8

Chapter 4   FRESHWATER ECOSYSTEMS.. 12

Chapter 5   AQUATIC PLANTS OF THE MEDIA LUNA WATERSHED 22

Chapter 6   FISHES OF THE MEDIA LUNA WATERSHED 40

Chapter 7   BIRDS OF THE MEDIA LUNA WATERSHED 55

Chapter 8   REPTILES AND AMPHIBIANS OF THE MEDIA LUNA WATERSHED 76

Chapter 9   CRUSTACEANS OF THE MEDIA LUNA WATERSHED 86

Chapter 10   ENVIRONMENTAL IMPACTS.. 89

GLOSSARY.. 97

SELECTED REFERENCES 101

List of Figures and Tables

Figure 1 - Los Ateojitos, a small aquatic ecosystem near Media Luna. Note the man-made "square" ponds where tilapia raising (aquaculture) was attempted. 2

Figure 2 - Los Peroles 3

Figure 3 - Prehistoric Media Luna. Many fossils of pre-historic animals and ancient civilizations have been recovered from Media Luna. Unfortunately, most artifacts have been looted. Illustration by Israel Ivan Garcia Martinez. 4

Figure 4 - These are the remains of early aqueducts that once provided Media Luna's water for human consumption and irrigation. A – Circa 1700's, B – circa early 1900's. 7

Figure 5 - The channels inside the park are favorite camping sites for visitors. Note the lack of any vegetation along the shore. 8

Figure 6 - Multiple use zones for Media Luna. Figure from the Official Journal of the State of San Luis Potosi, ANO L XXXVI S.L.P. 07 June 2003. Illustration by Israel Ivan Garcia Martinez. 9

Figure 7 - To the far left is the spring lake "Media Luna". To the right is the east channel that continues in a natural state for another 1.6 kilometers. 10

Figure 8 - A simple diagram showing the basics of the water cycle, also known as the hydrologic cycle. The constant movement of water on, overhead, and beneath the surface of the earth remains somewhat constant. Image by wawritto/ Shutterstock.com. 13

Figure 9 - Lentic zones. Illustration by Israel Ivan Garcia Martinez. 14

Figure 10 - Natural succession of a typical lake. Note how the sediment gradually builds up to form land. Illustration by Israel Ivan Garcia Martinez. 15

Figure 11 - A simple diagram of an aquatic food web. Illustration by Israel Ivan Garcia Martinez. 17

Figure 12 - Aquatic food chains begin with microscopic organisms and end with an apex predator. Illustration by Israel Ivan Garcia Martinez. 18

Figure 13 - A typical wetland showing the upland, riparian, and aquatic zones. Illustration by Israel Ivan Garcia Martinez. 20

Figure 14 - A representative wetland area east of Media Luna with thick patches of reeds. This small pond usually dries up during the summer. 21

Figure 15 - The usual way of describing aquatic vegetation. We have added algae to our description. Illustration by Israel Ivan Garcia Martinez.. 22

Figure 16 - Types of plant life found in an aquatic ecosystem. Illustration by Israel Ivan Garcia Martinez. 23

Figure 17 - Reed has many uses. It is used for fences and thatched structures throughout the entire watershed...31

Figure 18 - Note the triangular cross-section of most sedges................................... 33

Figure 19 - Small floating algae blooms observed in recent years. Photo by Ossiel Martinez....... 39

Figure 20 - Fire pits like this one cause severe damage to the natural environment. 93

Figure 21 - The irrigation canal that flows north of the springs becomes abundant with aquatic vegetation that supports thousands of crayfish, turtles, and fish................... 93

Figure 22 – Aquatic plants covered with silt. 94

Table 1   Guide for Differentiating Between Sedges, Grasses, and Rushes........................... 33

Table 2   Status of Fish Species Present in the Media Luna Watershed, Their Distribution and Extinction Risk Category According to the Mexican Official Standard - 059 (INE 2002) 41

*Listed are the figures used for the general descriptions of the Media Luna Watershed. Not listed are accurate figures/photos used for species identification in chapters 6 – 9.

We abuse land because we regard it as a commodity belonging to us. When we see land as a community to which we belong, we may begin to use it with love and respect. Aldo Leopold

Forward

In order to avoid confusión through translation, the following Forward from Professor Jorge Palacio Núñez will be provided in Spanish.

El manantial de la Media Luna es singular por varios motivos. Está formado por seis cráteres de manantial principales que, en conjunto, forman una laguna profunda. Además de la laguna, hay una serie de canales que surgen de ahí, que a su vez forman un sistema complejo de humedal que contrasta marcadamente en un entorno semiárido. De este conjunto de manantiales surge gran cantidad de agua termal de naturaleza sulfática cálcico-magnésica y extraordinariamente cristalina, lo que es casi una huella digital, pues pocos manantiales en el mundo presentan estas características.

Media Luna es sólo el mayor sistema de manantial de la región, y comparte varias peculiaridades con otros manantiales, tales como Anteojitos y Peroles. Todos ellos son ecosistemas relativamente aislados que surgieron en donde existió un gran lago que, biológicamente hablando, se mantuvo aislado de otros sistemas acuáticos. Ese aislamiento propició la aparición de al menos seis especies de peces endémicos para esta zona geográfica. Cuando este lago desapareció, los peces y otros endemismos acuáticos de la zona quedaron atrapados en el río Verde y en los otros manantiales y arroyos de la zona. Todos estos sitios son de gran belleza escénica, sobre todo en lo referente a sus singulares paisajes subacuáticos y a sus riberas. Sin embargo, debo mencionar cuatro singularidades más propias de Media Luna, que la hacen especial sobre los demás manantiales: 1) en tiempos prehistóricos fue parte de un yacimiento fosilífero, 2) en tiempos más recientes fue un centro ceremonial prehispánico, 3) comenzó a ser famosa para la ciencia desde principios de siglo XX, y todos estos endemismos de la zona fueron descritos aquí y, 4) presenta una extensión y una belleza que supera con creces a todos los demás manantiales de la zona, pero, paradójicamente, estas características han sido la causa inicial de su degradación biológica.

El agua de todos estos manantiales ha sido canalizada para irrigación agrícola; desde hace varios siglos en Media Luna, o desde tiempos más recientes en los otros manantiales. Con esto la biodiversidad pudo haber sido afectada, pero no hubo documentación de ello. En la actualidad, una fama creciente en el enfoque turístico ha estado atrayendo a numerosos visitantes que en Media Luna se contabilizan por decenas de miles anualmente. Este mismo enfoque también propició el interés de autoridades regionales y locales, así como de propietarios, a incorporarse a la industria del turismo. Por sí mismos, los turistas alteran las condiciones naturales de estos sitios, afectando a algunas especies nativas, sin embargo, son las acciones de manejo mal enfocado hacia la actividad turística lo que está afectando seriamente a estos singulares ecosistemas. Como resultado de esto, dos de las especies de peces endémicas de la zona ya desaparecieron en Media Luna y las demás se encuentran en mala situación poblacional. En Anteojitos se están siguiendo los mismos pasos y, aunque Peroles tiene cierta protección debida a su lejanía y aún no cuenta con infraestructura, también está creciendo en popularidad turística.

Muchos de estos aspectos negativos del turismo son debidos a falta de información. Existen trabajos derivados de la investigación científica donde se han documentado los efectos de diversas actividades, tanto de manejadores como de los propios turistas. Sin embargo, estos resultados son de baja difusión entre la población en general y suelen carecer de impacto.

Por esto, obras como las de Richard Worfel, donde, además de la descripción de estos ecosistemas, plantea tanto la problemática como la solución, son fundamentales para un entendimiento profundo por parte de los turistas y los manejadores.

Dr. Jorge Palacio Núñez
Profesor Investigador en Fauna Silvestre

Preface

From Richard:

I have been exploring the Rio Verde region for nearly twenty years with my good friend, Oceanographer and Master Scuba Diver Trainer, Ossiel Martinez. As a field biologist, I became frustrated with the lack of nature guidebooks for the Rio Verde region. I've always taken at least twenty pounds of nature reference books and guides of similar semiarid areas with hopes of finding the bird, plant, fish, snake, turtle or other critters that I observed in the Rio Verde area. Finally, I decided to make it simpler for people interested in learning about the region by composing a simple guide to the aquatic natural history of the Rio Verde area within the Media Luna Watershed.

The entire Media Luna Watershed is an uncommon territory due to its aquatic nature found within semiarid conditions (25 to 51 cm of annual precipitation). The Media Luna Watershed averages 50 cm per year. The entire watershed is a rare oasis of life surrounded by fields of citrus trees and various other crops: nourished by the waters from the Media Luna springs.

In this book, I provide a straightforward description of the Media Luna Watershed and what makes it such a valuable and unique ecosystem. I dedicate this book to all the foreign visitors and local citizens who are looking for their personal discoveries and adventures within the Media Luna Basin.

Ideally, the forces of nature should govern the Media Luna Watershed. But, this is not the case. Therefore, it is critical for us to band together, both political and scientific alliances, to preserve Media Luna as an environment where visitors feel safe, comfortable, and at home. Ultimately, we must learn to allow the watershed to live by its biological rules while allowing humans the opportunity to enjoy its many natural treasures.

During the many trips to the Rio Verde Watershed to gather information and photos for this book, I learned to admire the people of the region. Their friendship, kindness, and willingness to help me made this a terrific experience for a gringo from Texas.

Special thanks go to our wives, Sara and Erika. Sara provided unquestionable support during long periods while I was traveling to Rio Verde to collect information for this book. Erika kept us fed and made sure I always had a safe and comfortable place to stay. The authors are grateful to naturalist Fred Wills for his professional consultations and advice. I wish to acknowledge the encouragements and many excellent contributions of my friends. Dr. Jorge Palacio-Núñez's guidance with fish identification and technical information was especially important and valuable. Seth Patterson's professional ability to capture pictures of the fish in their natural habitat is greatly appreciated. The photos of the underwater world are from Seth unless otherwise noted. For outstanding photos taken by Seth during his visit to Media Luna check out: http://www.scubaboard.com/community/threads/trip-report-la-media-luna-san-luis-potosi-picture-heavy.502411/. Unless otherwise noted, the other photos were taken by me.

From Ossiel:

Having been born in Rio Verde, Media Luna is a unique location for me. It provides me a natural and positive energy. When scuba diving in Media Luna, you are surrounded by nature that permits one to reload and engage in the beauty. When I was a child, and I had to swim with my mother, I saw Media Luna as a vast expanse of water that is full of life. In addition to having a great time playing in Media Luna, I was always fearful yet curious about what lies beyond. Now, after many years, I still see Media Luna as a large body of water but with a particular sense of value and respect.

I met Richard and his wife, Sara, when they came to Media Luna to take a diving course. I was surprised by his interest in nature. As time went by, our perception and sensitivity for Media Luna grew little by little. Our interest in the site was growing, and we shared a special respect and admiration for the spring.

As I grew to love Media Luna, my need to know its nature intensified. To love such a fragile place requires one to know it. But to know it requires sensitivity and patience to study it and respect it. Overall, it needs a particular respect for nature to write about it.

Every day, I relive my first experience of wonder and excitement while swimming and diving here as a child as I teach my students the excitement of breathing underwater in Media Luna. I see a transformation of each of my students after diving in Media Luna as they change and develop a new view of nature.

Our society's demands for travel is growing each year, and Media Luna does not escape this pressure. Every year the park is visited by more and more people, especially on weekends and holidays. Each time the use and abuse of the park grows. The impacts on the fragile shores of mud and dirt increase as people walk along the shores or get in and out of the water. The aquatic plants are destroyed, and many of the plants are covered with silt that settles on the leaves, preventing light from reaching them. The popularity of Media Luna has its daily price as more use means more time needed for recovery. Not long ago, recovery time after Holy Week was only a few days, now it takes months. The human impact is that much greater.

Richard has written this book at the right time with the goal to take action for a substantial and permanent change. I have been teaching diving and conservation at Media Luna for 23 years and have confidence that learning about the environmental impacts on Media Luna has high value. Man must find a balance between preservation and use of Media Luna to preserve it for future generations. Hopefully, through this work, we will learn to see Media Luna with sensitive eyes and respect for living things both in and out of the water, including man himself.

Chapter 1

INTRODUCTION

"Any fool can know. The point is to understand."
Albert Einstein

A watershed is the area of land that drains rain or spring water into one location such as a stream, lake, pond, marsh, or wetland. They are some of the most productive and fascinating ecosystems in North America. Watersheds provide diverse ecological benefits including water purification, recreation, irrigation, and habitat for plants and wildlife. Unfortunately, the amounts of watersheds around the world are becoming more and more threatened. Used by man and animals since prehistoric times, the Media Luna Watershed has withstood human development, but today it may be at risk. The author and dedicated individuals have recognized the need for educating the community about protecting the remaining watershed surroundings in-and-around the Media Luna area. Through education, we can work to increase the understanding of the Media Luna Watershed and foster appreciation for its natural beauty. These pages will facilitate a sharing of ecological relationships and ideas usable in many educational settings. Specifically, this material should prove useful in teaching watershed ecology, environmental sciences, aquatic flora and fauna identification, and earth science aimed at the Media Luna Watershed.

Watersheds are vulnerable to loss or inadvertent destruction related to land development, agriculture, and recreation. These water bodies provide water for drinking, agriculture, manufacturing, recreation, and habitat for numerous plants and animals. Only by knowing the Media Luna Watershed and its ecosystem can we begin to understand why and where small changes can have enormous impacts.

Included is a guide to the identification of the fishes, plants, and other critters that depend on the watershed. We will briefly cover the changing uses of Media Luna and the management system that is in place. We conclude with a discussion of the environmental impacts of Media Luna.

The discussions will cover the entire Media Luna Watershed although when the name, "Media Luna" is used by itself, we are only talking about the spring lake. Also, when we speak of the canals and channels; canals will mean man-made waterways while channels will be natural waterways.

This book is a simple guide for those interested in learning more about the natural history of the Media Luna Watershed. It is not all-inclusive of the aquatic plants and birds seen around the watershed. The objective is to identify the most common aquatic plants and birds that depend on the watershed. The chapter on the fishes of the Media Luna Watershed is current. The overall goal is to provide a learning tool for the new or amateur naturalist or biologist, as well as any person with a curiosity about what life exists within the watershed. We have incorporated Laguna Los Peroles and A la Ateojitos into the territory covered because of their biological and geologic links to Media Luna.

Los Anteojitos is a smaller aquatic habitat with two modest lagoons connected by a small channel. Los Anteojitos is only 6 km from Media Luna and believed to have subterranean connections. The overexploited Los Anteojitos is now seldom used by tourists.

Los Peroles is another aquatic habitat with possible subterranean connections. It is smaller but similar to Media Luna. Some people call it, "Little Media Luna." It has a depth of nearly 9.144 meters and has many of the same species of flora and fish as Media Luna. Both water sources provide valuable water to surrounding farms through extensive irrigation systems. An interesting fact about Los Peroles is that the oldest known trees in Mexico are in the wetland that includes Los Peroles. These astonishing Montezuma Bald Cypress (*Taxodium mucronatum*) consist of trees over 1000 years old. The largest individuals at Los Peroles are more than 22 meters tall and 2 meters in diameter.

Figure 1 - Los Ateojitos, a small aquatic ecosystem near Media Luna. Note the man-made "square" ponds where tilapia raising (aquaculture) was attempted.

Figure 2 - Los Peroles

DESCRIPTION OF THE MEDIA LUNA SYSTEM

"In the end, we will conserve only what we love; we will love only what we understand, and we will understand only what we are taught"-Baba Dioum

La Media Luna Watershed is in the Rio Verde Valley (Region Medio) within of the State of San Luis Potosi. The watershed is in the high semiarid plateau of the Sierra Madre Oriental Mountains. The watershed is rich in agriculture and adjacent to the Huasteca Region, a region known for outstanding rafting, rock climbing, kayaking, and exploring. The whole basin (watershed) is called the Rio Verde Valley (Green River Valley).

In prehistoric times, the valley was a vast marine lake covering the entire plateau. The marine lake resulted in the formation of limestone sedimentary rocks, formed by the mineral calcite that came from the beds of evaporated lakes and sea animal shells.

Fossil records found in-and-around Media Luna reveals that mastodons and other prehistoric creatures were frequent visitors to the clear waters of La Media Luna. Based on artifacts discovered in the lake, it is possible that ancient indigenous ceremonies occurred as early as 800 BCE. Old cypress trees that fell into the lake lie on the lake's floor. The locals like to refer to then as the "petrified forest" but petrification is probably not complete. Instead, they are in conditions called, "permineralization and replacement." These two processes produce copies of the original specimen, so the cedar logs in Media Luna are in the course of becoming truly petrified. This process requires more than a minimum of 10,000 years to take place.

Figure 3 - Prehistoric Media Luna. Many fossils of pre-historic animals and ancient civilizations have been recovered from Media Luna. Unfortunately, most artifacts have been looted. Illustration by Israel Ivan Garcia Martinez.

Using fluorescent tracers, scientists have determined that the water in the aquifer that supplies Media Luna originates in the mountains that surround the Rio Verde Basin. Fluorescent uranium tracers are added to standing water after a rainfall before the water soaks into the ground. Rainwater over the mountains and in the basin percolates through the underground water systems and eventually ends up in various springs throughout the region. Water is then collected from springs to determine if the tracer is present. Using fluorescent tracers is an effective way to learn about how a water cycle works in a chosen area. The primary water cycle for Media Luna's waters begins over the Gulf of Mexico through evaporation that blows over the land. The process of transpiration adds moisture to the atmosphere. The moist air blows over the mountains where precipitation occurs. The rainwater runoff percolates into the ground to feed the subsurface water system. Some of the rainwater will evaporate and result in precipitation over the basin. The groundwater escapes to the earth's surface in the springs of Media Luna that forms a lake. Small amounts of water evaporate from the lake. Most of the water is tapped into the irrigation system. Water not used by plants or other purposes flows into the Rio Verde and eventually back into the Gulf of Mexico.

The speed of underground water movements (i.e. from the mountains to Media Luna) can be amazingly slow-moving; it may take rainwater tens or even hundreds of years to reach the aquifer it is filling and finally to a spring.

Engineers have mechanically attempted tapping the fresh water from the lake, but each attempt has failed. Recently, the concrete pedestals once used to support massive pipes that carried pumped water from the lake were removed. The overflow is channeled by gravitation to more than 15,000 hectares of farmland for irrigation thus changing a mountain desert-land into a citrus fruit and vegetable bonanza. The Media Luna Watershed is of significant economic importance to the surrounding communities due to both its irrigation and recreational activities.

Gravity and topography define a watershed. Most watersheds drain into a marsh, stream, river or seep into the ground. Gravity and topography move the water into irrigation channels and ditches and seeps into the earth to nourish crops. The area of land surrounding the network of lakes drains into the watershed. Sometimes, in the case with the Media Luna Watershed, the gravity and topography are manipulated to stimulate the flow of the irrigation water. Ultimately, the water merges into another river, "Rio Santa Maria," that empties into the Panuco River. The Panuco River finally flows into the Gulf of Mexico.

Unknown to many, but is particularly important when talking about the entire Media Luna Watershed, are two others spring lakes, one called "Los Peroles" and the other called "Anteojitos." Nearby marshlands consist of an unknown number of seeps (where water oozes from the ground through surface sands). Thirty-eight km north/northeast is the small lake "Los Peroles" with three main springs and a depth of around 11 meters. Both of these systems obtain their water from the same aquifer source as Media Luna. For the purpose of this book, Los Peroles and Anteojitos are considered part of the Media Luna Watershed.

One can observe how topography plays a role in a watershed by visiting the Rio Verde River a few miles west of the town of Rio Verde (see figure 5) where the steep hillsides/banks along the river controls the river flows (riparian area). It is easy to see that the topography has dropped thus allowing gravity to move the remaining waters from Media Luna to flow into the river. At this point, we have a new and bigger watershed known as the Rio Verde. It is like the blood's circulatory system in the body. The smaller capillaries feed into small veins that eventually flow into the cardinal veins and ends up at the heart (or regarding a watershed the water ends up in the ocean).

The sulfurous water in the Media Luna Watershed comes from limestone karst springs. The slight smell of sulfur is a product of decomposing organic matter that produces hydrogen sulfide. Karst springs form when naturally acidic water seeps into the limestone forming underground channels and caves that hold and transport water. Media Luna is an artesian spring because the natural pressure is high enough to reach the surface. What we see at Media Luna are the ends of a limestone aquifer system where a cave reaches the earth's surface. The springs are the origin "headwaters" of the Media Luna Watershed.

Aquifers are beds or layers of permeable rock (sometimes forming large cave systems). Artesian aquifers flow through and out the rock formations to springs; such is the situation at the karst springs at Media Luna. Underground aquifers supply water for wells, springs, lakes, and rivers. The El Doctor aquifer provides water for the Media Luna Watershed.

Media Luna's artesian springs come in two general types, karst springs (where water pours out from caves dissolved from a limestone aquifer) and seeps (where water from an aquifer slowly reaches the earth's surface). The majority of Media Luna's Watershed comes from karsts springs although there are numerous seeps in and around the lake. Total exchange of Media Luna lake water with new spring water takes place an estimated 24 – 36 hours, making it a unique ecosystem. There are six key springs in Media Luna at depths of 10 to 38 meters with a flow rate of 4.35m3/sec.

A spring in the Media Luna system is an area where the aquifer reaches the earth's surface. The water may slowly trickle out and is sometimes never noticed, or it can come out with an immense force of the major springs at Media Luna. Here, the pressurized water in the aquifer pushes upward to the springs through fissures and fractures in the karst structure. The reason the spring water is warm (26 - 28 degrees Centigrade) may be due to molten rock deep in the earth or a chemical reaction occurring in the aquifer/karst system – or a result of both.

The springs create select ecosystems for flora and fauna in an area that is naturally a dry semiarid region. The lake, channels, and canals support shared and endemic species of plants and fish. One can often find turtles sunning themselves and various wading birds hunting for a meal or roosting in a tree's canopy along the shores.

The springs have played a crucial role in supporting prehistoric animals and indigenous tribes. Media Luna has been a storehouse of objects made by ancient people as well as fossils of extinct animals, such as the mastodon. One of the unique features of Media Luna is the fossil woods found in the lake. Huge cypress trees sunk into the lake hundreds, maybe thousands of years ago and are preserved by exceptionally low oxygen levels.

NOTE: It is illegal to move or remove artifacts from the Media Luna system without permission from the proper governmental authorities. There is an excellent collection of fossils and artifacts found near Media Luna at Museo Regional del Rio Verde, Located at Mariano Matamoros 143, Centro, CP 79616, Rio Verde, SLP – Tels: (487) 872-3224. Unfortunately, most of the artifacts have been looted and were taken to the USA or are in the hands of a local residents. A small museum is inside the park. Readers are encouraged to visit these two locations for a genuine appreciation of the region's history.

A

B

Figure 4 - These are the remains of early aqueducts that once provided Media Luna's water for human consumption and irrigation. A – Circa 1700's, B – circa early 1900's.

Media Luna has a written history that goes back to 1617 when the erections of two aqueducts to the village of Rio Verde were constructed. An irrigation ditch Built in 1732 carried water to irrigate Fernandez for the growing of sugar cane. In 1917, a new enlarged irrigation system was needed to handle the introduction of citrus fruits that are today's leading crop. Again, in 1977, the Media Luna irrigation system expanded to develop and enlarge the irrigated areas. One of the most noteworthy roles the Media Luna Watershed plays today is with its irrigation system. Corn, oranges, sweet potatoes, sugarcane, and beans grow all year long. They provide a vital role in the region's economy.

THE CHANGING ENVIRONMENT OF MEDIA LUNA

We must all obey the great law of change. It is the most powerful law of nature.
Edmund Burke

Media Luna has undergone a complete facelift over the last twelve years. It has developed into a popular tourist destination, drawn by the fresh, clear waters of the Media Luna spring system. In 2003, Media Luna became a State Park and protected natural area. Operated and managed by the local Ejido system, El Jabali, Media Luna has become a gold mine for the Ejido. The Ejido system is a government that promotes the use of communal land. Everyone in the community shares the area. The Ejido leadership is made up of local citizens, mostly farmers, who take turns managing the Ejido. As many as 6,000 daily visitors visit the park over the holidays where they spend thousands of pesos on entrance fees, food, water toys and drinks at booths operated by citizens of the Ejido. On an average weekend day, there are six or more parked buses at the entrance as well as over 200 cars in the parking lot. The park has quickly become a star attraction for people looking for an inexpensive means of family outdoor recreation.

Figure 5 - The channels inside the park are favorite camping sites for visitors. Note the lack of any vegetation along the shore.

Some of the quaintness and mystique of Media Luna are lost among the crowds, although the scuba diving is still extraordinary. Media Luna has become a State Park and protected natural area. Today Media Luna is "Parque Estatal de la Media Luna."

Parque Estatal de la Media Luna has an environmental management plan established by the State of San Luis Potosi, Secretariat of Ecology and Environmental Management. It divides the park land into distinct zones, each with an assigned use. The multiple use concept is supposed to protect the most environmentally sensitive regions. Although, strict enforcement of the rules is essential to the environmental management plan to succeed.

The following are the seven multiple use zones:

1. Zone of Conservation, Scientific and Research Use.
2. Productive use, specifically farming.
3. Productive Use, specifically aquaculture.
4. Productive use, farming in chinampa pilot area.
5. Tourism.
6. The area in Recovery. Tourist Use Restricted.
7. Buffer Area, Restricted Use.

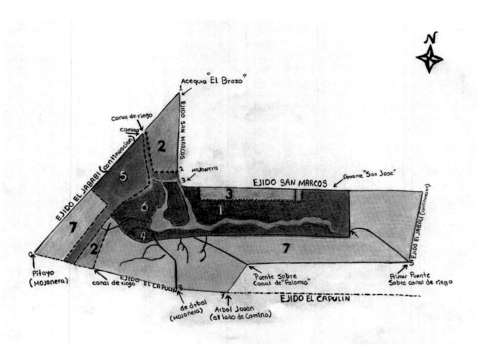

Figure 6 - Multiple use zones for Media Luna. Figure from the Official Journal of the State of San Luis Potosi, ANO L XXXVI S.L.P. 07 June 2003. Illustration by Israel Ivan Garcia Martinez.

It is important to note that zones 1 and 6 are the two zones most critical for the ecological maintenance of aquatic flora and fauna. Zone 6 has undergone many damaging environmental impacts over the years. Due to the many non-native Australian Pines in this zone, recovery to a natural state may never occur.

Figure 7 - To the far left is the spring lake "Media Luna". To the right is the east channel that continues in a natural state for another 1.6 kilometers.

There is a short strip of paved road leading to Media Luna from Highway 70. For the most part, the rest of the drive is still dusty and bumpy. On wet days, the road is muddy and can be slippery in spots. One does not require a four-wheel drive vehicle to get to Media Luna. The drive to the lake from Highway 70 has not changed much in the last 20 years. The road to Media Luna is scheduled for paving shortly. One of the main irrigation channels (North Channel) runs directly parallel to the road. The irrigation canals water's edge is only about two meters from the rim of the road.

A recent change has been the fencing-off of the entire park area. This action has worked well to keep freeloaders from entering the park. Ejito employees patrol the fence line. Therefore, the visitors must now enter from a single entrance for a small fee. It used to be free. The basic entrance fee is only 30 pesos per day. If one wishes to scuba dive, the fee is 70 pesos. The camping fee is 70 pesos per day (these prices are current at the time of writing this book.)

It is no longer possible to drive to the boardwalk that leads to the main dock. Now, one must park in the parking lot and carry items a few hundred meters or so to the lake. Automobiles used to pull up to the channels edge; allowing oil and other fluids to drip from engines and seep into the water system. The new parking area is an active effort to help keep the water free of pollutants.

New sidewalks and boardwalks lead to the diving and swimming docks. The sidewalks are an excellent improvement, especially appreciated on rainy days. New sales-booths, made of concrete blocks, have replaced the old thatched-roof shacks that sold goods to visitors. The old thatched roof huts were quaint looking and ideal for capturing interesting photos, but liquid waste could always be seen forming films along the shores of the channels. There are over fifteen shops that sell just about everything one needs while picnicking. Tacos, beer, and hot meals are a few of the most popular items. Improved drainage benefits the water quality in the lake and channels. Waste is no longer dumped onto the open ground where it could leach into the underground water system. Of course, the sanitation has also improved as concrete floors replace the old dirt floors. There are new restrooms adjacent to the sales booths with a controlled drainage (captured) system that no longer puts the sewage into the channels.

A new shoreline boardwalk runs adjacently to where there used to be campsites. The boardwalk is sturdy and convenient for swimmers and divers. It connects to the main expanded shoreline dock. A floating dock for swimmers lies securely anchored at about fifty meters from the main dock. A suspended rope lies stretched between the main dock and the opposite end of the lake. Swimmers use the line as a safety guideline.

Several cabins now exist in the park. The cabins are rented out to tourists. They sit about 200 meters from the lake.

A new camping area lies along the irrigation canal near the entrance. People used to camp under the huge evergreen trees next to the lake. The idea of moving the camping to outside the tree canopy was to give the understory a chance to grow back to its natural state.

The biggest change has been the management of the park area. There never was consistent control before becoming a state park. Before, there were few visitors, and most everyone seemed to appreciate and respect the natural surroundings. Today, there are numerous tourists (up to 6,000 during holidays such as Easter), and compliance with the natural environment is questionable. Part of this appears to be a lack of control and enforcement of the rules established by the State of San Luis Potosi.

An increased amount of construction (cottages, permanent homes, rental cabins, and businesses) is taking place along the watershed's canals. Most of this is due to the boost in tourism at Media Luna. The new construction has a negative impact on the unique flora and fauna along the canals.

Chapter 4

FRESHWATER ECOSYSTEMS

We forget that the water cycle and the life cycle are one.
Jacques Cousteau

THE WATER CYCLE

The water cycle, (also known as the hydrologic cycle) that feeds Media Luna is always in constant movement on, under, and above the basin surface. Water in the water cycle is always changing between liquid and vapor states. Driven by energy from the sun and power of gravity, life-sustaining water finds its way to the Media Luna Watershed. There is no beginning or ending to the water cycle; it continues working to move water from one location and state to another.

The major components of the water cycle are evaporation, condensation, transpiration, precipitation, and runoff. Evaporation is where the sun heats surface water to form water vapor. In Media Luna's situation, most evaporation occurs in the Gulf of Mexico. Evaporation of another kind takes place through plants and trees where moisture evaporates into water vapor (this system is called transpiration).

The water vapor then condenses (condensation) returning the gas back into a liquid state. The condensation may change to precipitation (rain) over the Gulf or blown over land to form clouds, mist, fog, dew, or frost.

The clouds are blown by the prevailing winds across the land until they become too heavy (usually in the area of the mountains), and they fall to the earth as precipitation (rain). Plants then absorb the water or filtered through the soils and into underground aquifers. The precipitation can also evaporate and form new clouds. Water can run off and flow into rivers, streams, marshes, bogs, swamps or lakes where it will eventually seep into the groundwater, evaporate, or flow back to the Gulf. Regardless to the pathway precipitation takes, it will always continue to cycle or recycle back into the water cycle.

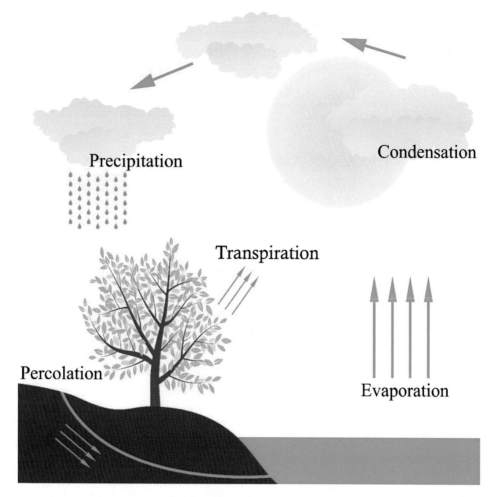

Figure 8 - A simple diagram showing the basics of the water cycle, also known as the hydrologic cycle. The constant movement of water on, overhead, and beneath the surface of the earth remains somewhat constant. Image by wawritto/Shutterstock.com.

Freshwater ecosystems are water systems having a low salt concentration (usually less than 1%). Flora and fauna that live in freshwater have adjusted to low salt conditions and are not able to survive in saltwater. Freshwater ecosystems consist of ponds, rivers/streams, marshes, and lakes; all found in the Media Luna Watershed.

LENTIC AND LOTIC ECOSYSTEMS

Freshwater ecosystems fall into two distinct categories: **lotic** – meaning running water found in rivers and streams and **lentic** – meaning standing water such as lakes, ponds, marshes, and swamps. The Media Luna Watershed is made up of both categories.

Lentic ecosystems have well-defined borders; the water's edge, the sides of the basin, the surface of the water and the floor sediments. Three zones make up a lentic ecosystem.

1. Littoral zone – a shallow vegetated zone

2. Limnetic zone – an area that light penetrates in the open water
3. Profundal zone – the area where the bottom sediments are with no light penetration

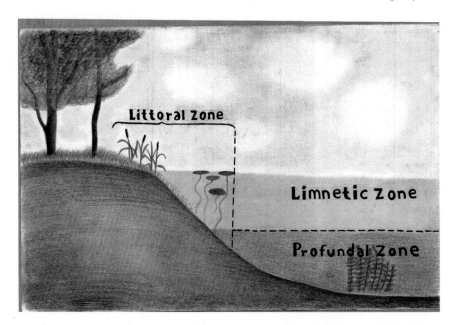

Figure 9 - Lentic zones. Illustration by Israel Ivan Garcia Martinez.

The littoral zone is shallow and where most aquatic plants grow. This area is easy to see at Media Luna; it is where one can see the broad leaves of the lily pads. This area is shallow enough for rooted aquatic plants to receive plenty of sunlight. In Media Luna, the plants are rooted in the silt and clay. The littoral zone provides habitat for a variety of animals such as fish and turtles. The littoral zone is also the area where most aquatic insects subsist. There are four types of aquatic plants in the littoral area: emergent, floating-leaf, submerged and algae. Littoral plants are discussed in Chapter 5, "AQUATIC PLANTS OF THE MEDIA LUNA WATERSHED." Algae is one of the few plants that can survive in the profundal zone.

The open water of a lake is called the limnetic zone and is the area of no rooted plants and limited light penetration. This area is inhabited by fish and plankton but no plant life other than some plant-like algae. The amount of light penetration will depend on the amount of suspended solids in the water. In Media Luna, the suspended primary particles are silt from the bottom of the lake as well as organic and inorganic particles from the shoreline. Much of the limnetic zone is without any aquatic vegetation due to the tremendous layers of silt, clay, and detritus and, of course, the lack of adequate sunlight.

The profundal zone is at the bottom of a lake where little if any, light penetrates. The bottom is a mixture of silt and clay with some organic matter mixed in. Media Luna has an extensive profundal zone where nothing will grow, although some fish, crayfish, and turtles may inhabit this region.

The lotic ecosystem is distinguished by running water and currents. Lotic systems have unidirectional gravity influenced current. Lotic environments fall into two different zones:

- Lotic erosion: fast running water with riffles where erosion can take place
- Lotic depositional: slower velocities of water where erosion is not expected

Lotic erosion areas are along the shores and banks of the canals. Lotic depositional zones are along the natural channels. Water flow velocity in the Media Luna Watershed is controlled by manmade water gates. The water gates ensure water is always flowing and puts a man in total control.

Lotic ecosystems may be as large as major rivers that flow for thousands of miles to tiny seeps (springs) where only a few cm of water flows per hour. Frequently, lotic systems have segments that are much like lentic systems. These segments have slow moving water due to their depth. There is more biotic diversity within lotic systems because numerous living organisms must be specialized to exist in flowing water situations. Both lotic and lentic ecosystems have light, chemistry, temperature, and different substrate materials that play important roles in determining what types of organisms that live in them. Only lotic systems have "flow" as a key determinant.

Most lakes go through a natural aging process known as, "succession." Organic and inorganic sediment gradually builds up to a point that a lake matures into a pond, swamp, or marsh and eventually into a meadow. Throughout various stages of succession, the types of plants and animals in a lake community change. It takes many hundreds of years for succession to complete. Media Luna Lake is unique in that it does not follow the usual succession model for lakes. The uniqueness is due to three major factors:

1. Constant rapid inflow of new water through the subsurface springs
2. Unlike temperate zones, arid conditions do not support vegetative growth along the banks
3. No streams or rivers flow into the lake

Figure 10 - Natural succession of a typical lake. Note how the sediment gradually builds up to form land. Illustration by Israel Ivan Garcia Martinez.

Does the flow of water created by the massive springs make Media Luna a lotic system? Not exactly, but it is the primary factor that makes the lake unique. So, for the purpose of this guide, we will consider the lake as a lentic system.

FOOD CHAINS AND FOOD WEBS

All living critters require food for energy to grow, reproduce, and move. Food chains show how each living critter gets its food. It shows who eats who. Food chains start with autotrophs such as green plants and small microbes that obtain their energy from the sun. Some scientists call them producers. A food chain ends with apex predators that are called consumers and are at the top of the food web with few or no predators of its own. A food web consists of multiple food chains within a single ecosystem. Therefore, a food web is a complicated system consisting of all the unified and overlapping food chains.

Living organisms in a food web form groups called trophic levels with the bottom trophic level being producers and the top being the consumers. Most autotrophs produce glucose (energy) through a chemical process between sunlight, carbon dioxide, and water – this process is called photosynthesis. They also use nutrients in their surrounding environment. In an aquatic environment, the most familiar Autotrophs are plants, but algae, phytoplankton, and some bacteria are also autotrophs.

Consumers are the animals that eat producers. Herbivores (plant eaters) consume the plants, algae, phytoplankton, and some bacteria and are primary consumers. Thus the next trophic level. Secondary consumers eat the herbivores and are the third trophic level. The tertiary consumers are the next level (fourth), and they eat the secondary consumers. Depending on the ecosystem, this system can go on and on until it reaches the apex predators.

The primary source of energy in an aquatic ecosystem is solar energy. Solar energy is captured by tiny producer organisms called phytoplankton (free-floating aquatic plants). Zooplanktons (free floating aquatic animals) are primary consumers and consume the phytoplankton. Secondary consumers, such as macroinvertebrates and some fish, consume the zooplanktons. Larger fish then consume the secondary consumers. Finally, we come to the apex predator that consumes the larger fish.

Consumers that eat both animals and plants are called omnivores. In the human environment, people are omnivores because we eat both plants and animals.

The last part of a food web (or the first part depending on how one looks at it) is made up of detritivores and decomposers. They consume dead plant and animal remains. Decomposers turn organic wastes into nutrient-rich materials thus completing the life cycle. At this point, the whole cycle begins again as the autotrophs use the nutrient material and sunlight for energy.

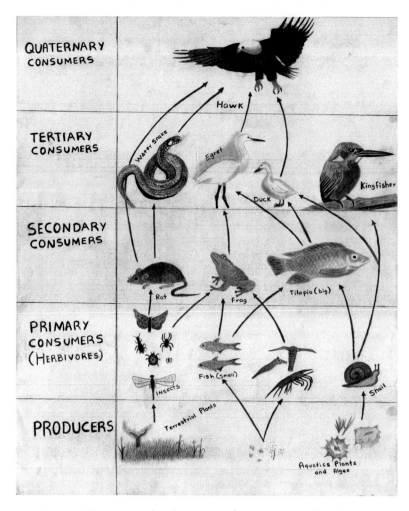

QUATERNARY CONSUMERS

TERTIARY CONSUMERS

SECONDARY CONSUMERS

PRIMARY CONSUMERS (HERBIVORES)

PRODUCERS

Hawk

Water Snake Egret Duck Kingfisher

Rat Frog Tilapia (big)

Insects Fish (small) Snail

Terrestrial Plants Aquatics Plants and Algae

Figure 11 - A simple diagram of an aquatic food web.
Illustration by Israel Ivan Garcia Martinez.

NOTE: Every scuba diver or swimmer in Media Luna has noticed the silt that arises from the bottom of the lake when touched. That is primarily a result of the detritivores and decomposers doing their work.

So why is this information necessary for a book about the Media Luna Watershed? The answer is that every link in the food web is related to at least two other links. The quality of an ecosystem depends on the balance of all these links. If one of the links is in danger, the entire food web is in jeopardy. The quality of the ecosystem will decline if only one phase of the food web link is under stress. Any change can result in a species' shift, a change in the distribution of a species

Figure 12 - Aquatic food chains begin with microscopic organisms and end with an apex predator. Illustration by Israel Ivan Garcia Martinez.

For example, the loss of plants will lead to a decrease in the plant eater's population. The decline in plant eaters will affect the next trophic level and so on up the web. The ecosystem is then out of balance. When this happens, we may see a decrease in the biodiversity of the ecosystem.

For this book, aquatic habitat refers to the natural home for plants and animals within the Media Luna Watershed. Often, people only consider the habitat around the lake as Media Luna's aquatic habitat, but that is only a small portion of it. The majority of the aquatic habitat is around the channels, canals, and irrigation ditches that receive water from the spring lake. A high-quality habitat includes cover, food, water, and a place to bring up young. A healthy aquatic habitat is critical for the survival of all plants and animals and the overall use of the watershed. The most critical habitat within the Watershed is the east channel where extra protection should be implemented.

WATER QUALITY

Water quality plays a significant role in determining the health of the aquatic habitat. Water quality is a term used to depict the physical, biological, and chemical characteristics of the water. These features are used to determine if the water is suitable for drinking, irrigation, recreation, wildlife or other desired use. Standard tests for water quality include:

- pH
- electrical conductivity
- fecal bacteria
- nutrients
- dissolved oxygen
- temperature

- total suspended solids
- turbidity
- discharge
- microinvertebrates
- biological oxygen demand
- submerged aquatic vegetation

An explanation of each of these tests is beyond the scope of this book, although, we will discuss three of the tests and their association with the Media Luna Watershed.

1. Fecal bacteria are in human and other animal's fecal matter. Not all fecal bacteria are harmful to humans. Those that are dangerous require several days in the laboratory for identification. Therefore, a test called, "total bacteria count" is used as an indicator of water's quality and safety because it is rapid (closer to real time). When the total numbers of bacteria are high, there is a higher probability of a harmful bacteria being present. With the greater amounts of bacteria comes a greater chance of suffering from ear infections, dysentery, typhoid fever, or gastroenteritis. Methods of contamination at Media Luna Lake are pets, livestock, wildlife, humans, (especially children without proper diapers), and runoff from human activity.
On occasions, like Easter Holiday, when over 6,000 visitors attend events at Media Luna, the chances of coming in contact with infectious bacteria are much greater for a swimmer or scuba diver. Fortunately, due to the high-velocity springs at the bottom of the lake, the freshwater replaces the infested water rather quickly. Although, this could result in highly contaminated water further down the watershed.

2. Dissolved oxygen is the amount of dissolved gaseous oxygen in the water. Gaseous oxygen enters the water through many ways. Most gaseous oxygen enters Media Luna Watershed water by diffusion, the passive transfer of molecules or particles along a concentration gradient, or from regions of higher to lower levels. Dissolved oxygen also enters the water through agitation and fast movement thus resulting in aeration. Dissolved oxygen is also a by-product of photosynthesis. Dissolved oxygen is removed from the water by respiration and decomposition of organic matter. Therefore, excessive dead vegetation, to include algae, result in larger quantities of organic matter and can deplete dissolved oxygen.
Dissolved oxygen is a good indicator of a healthy water system. Dissolved oxygen supports the aquatic life of all kinds. After all, fish breathe by absorbing dissolved oxygen from the water through their gills. Low dissolved oxygen levels put stress on all living things in a watershed. A body of water with high levels of dissolved oxygen has the potential for superior amounts of plants and animals.

3. Total suspended solids are solids that float on the surface or become suspended in the water. These solids can be made up of silt, decaying plant and animal matter, wastes, and sewage. Suspended solids do not usually include algae and phytoplankton.

Total suspended solids are closely related to turbidity. Scuba divers experience the result of high levels of suspended solids by seeing a decrease in visibility with an increase in suspended solids. Suspended solids can significantly reduce the amount of sunlight penetrating a body of water by absorbing, reflecting, and refracting the sunlight. The result can have a tremendous consequence on aquatic plants that require sunlight for growth and reproduction.

Temperature varies in most lake systems, both seasonally and by the depth (colder as one goes deeper). In between two temperature layers is a thin layer called the thermocline, the point where temperatures rapidly change. This process of temperature layers is "thermal stratification."

Typical thermal stratification of lakes does not apply to Media Luna. Media Luna is a thermal lake fed by bottom springs where the water enters the lake at near 86° C and cools off to around 82° C at the surface. Therefore, the stratification is the opposite for a thermal lake such as Media Luna. Warm water holds a lesser amount of oxygen than cold water so that it may become saturated with oxygen but still not contain enough for the survival of some aquatic life. Also, some compounds are extra toxic to aquatic life at elevated temperatures. For these reasons, Media Luna provides a unique habitat for specialized flora and fauna.

A discussion about freshwater ecology would not be complete without a description of a wetland. Wetlands exists where saturation with water is the dominant factor determining the nature of soil development and the types of plant and animal communities living in the ground and on its surface (Cowardin, December 1979). This habitat supports many animals, birds, amphibians, reptiles, some fish and a variety of plants. Many wetlands are seasonal. Wetlands are swamps, marshes, bogs and similar areas.

Just south of Media Luna is a hidden wetland. Other than a few local farmers, hardly anyone knows of its existence. These wetlands are a bio-diverse habitat in the Media Luna freshwater system. Perhaps, one may find a greater diversity of plants and animals living in and around Media Luna's sister wetland than around the lake or its canals.

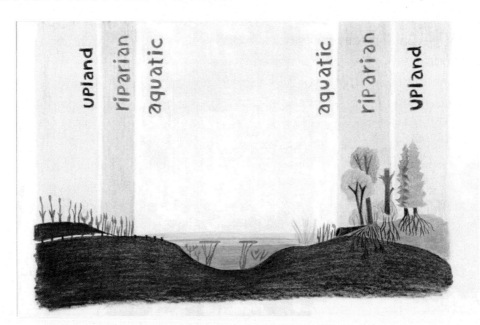

Figure 13 - A typical wetland showing the upland, riparian, and aquatic zones. Illustration by Israel Ivan Garcia Martinez.

Wetlands do not have to be wet all the time. Although, most have standing water or damp soils, at least, some part of the year. Wetland loss has been significant in Mexico since the 1700's due to agriculture, mosquito control, water diversion projects, and urbanization. At one time, wetlands were viewed as wastelands throughout North America. As we have expanded our knowledge about wetlands, we have learned there is a fragile interrelationship between wetlands, humans, and a healthy natural environment.

Wetland soils and plants naturally collect and filter nutrients and sediments. The slow and calm characteristic of wetlands allows for the settling out of materials and gradual seepage of naturally filtered water into the underground water column allowing for a generous number of plants in the wetland to take up particular nutrients from the water. As a result, we have cleaner groundwater and greater variety of vegetation.

Figure 14 - A representative wetland area east of Media Luna with thick patches of reeds. This small pond usually dries up during the summer.

Chapter 5

AQUATIC PLANTS OF THE MEDIA LUNA WATERSHED

A weed is a plant whose virtue is not yet known.
Ralph Waldo Emerson (1803 - 1882)

For the purpose of this guide, we will focus on the aquatic vegetation identified during surveys done from 2013 and 2014 of the Media Luna Watershed. The list includes the most common species of aquatic plants observed during our studies. Aquatic plants are plants growing partially or entirely in water whether fixed in the waters bottom or suspended without anchorage. Listed are each plant's common name followed by its scientific name using binomial nomenclature (genus and species). The natural vegetation of the area has received minimal scientific study resulting in inadequate technical literature. Our surveys included areas within the Media Luna Watershed to include the lakes, channels, and canals.

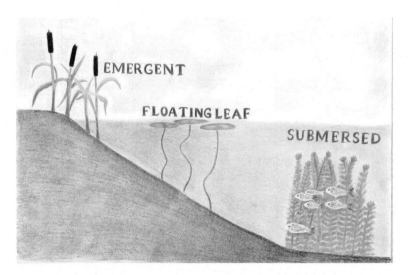

Figure 15 - The usual way of describing aquatic vegetation. We have added algae to our description. Illustration by Israel Ivan Garcia Martinez.

For this book, aquatic vegetation is divided into four groups:

- Floating Leaf Plants
- Submerged Plants
- Emergent Plants
- Algae

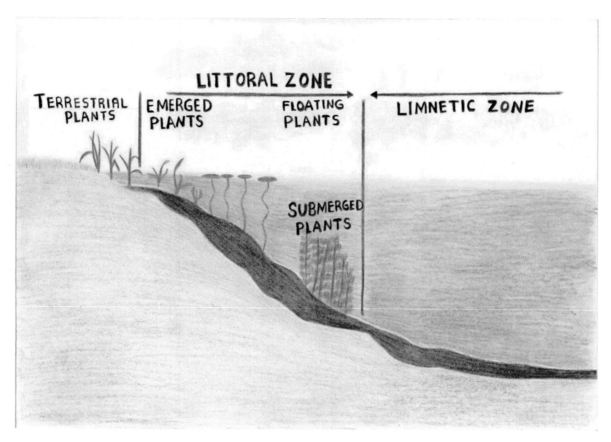

Figure 16 - Types of plant life found in an aquatic ecosystem.
Illustration by Israel Ivan Garcia Martinez.

Underwater sections of aquatic plants offer habitats for numerous micro and macro invertebrates. The invertebrates are utilized as food by many species of fish throughout the watershed. Amphibians, birds, and reptiles also consume the invertebrates. Even after aquatic plants expire, they are decomposed by fungi and bacteria and develop detritus, a valuable food for aquatic invertebrates. Aquatic plants play a vital role in maintaining a healthy aquatic environment. Many aquatic plants are beneficial towards filtering nutrients and toxic chemicals as well as stabilizing shorelines. Aquatic plants are an important fish habitat throughout the watershed.

One cannot discuss the plants of Media Luna without mentioning one of the most outstanding features of the lake. Scuba divers know about the hidden cave-like feature that can only be entered from underwater. It is formed by the massive roots of plants that grow over underground the tunnels.

Floating Leaf Plants

Floating leaf plants are of two types: those firmly rooted with floating leaves and those not anchored to the floor of the water system. Many of them have floating leaves with hanging roots that never touch the floor of the water system. Some floating plants are tiny, such as duckweed while others may be over a foot in diameter. No non-rooted floating leaf plants were found during our surveys. One primary reason for this is that floating plants prefer standing "still" water and almost all of the water in the watershed is continuously flowing.

Fragrant Water Lily
Nymphaea odorata lilies
Other names: Beaver Root, American white waterlily, Sweetwater lily
Spanish: Ninfea Blanco

The major floating-leaf plant found is *Nymphaea odorata lilies* (water lilies), a plant firmly rooted to the floor of the water system. The Fragrant Water Lily is a partially submerged plant with floating leaves. The heart-shaped to round leaves range in size from 10-30 cm wide. The underside of the leave is purplish-red with many veins. Their leaves are tough because they have to resist the weather and water movement. The round stems are rigid and attached to a spongy rhizome root-base buried in the sediments. The flowers are white and range from 7.5 – 12.5 cm wide. The surface leaves are nearly circular. There is a deep cleft where the stem attaches to the center of the leaf. Their white petals are curved lengthwise and form a small channel. Bright yellow stamens are densely packed in the center of the flower where there is a single pistil. They prefer shallow still water 1.5 – 2.2 meters deep. Although, the Fragrant Water Lily grows in water up to 7.62 meters deep in Media Luna. Possibly due to the crystal clear water in the spring. They favor bodies of water with silty or muddy bottoms. Water Lilies are a hardy plant that 's hard to destroy, mainly because they reproduce by seed and from their large spreading roots called rhizomes. A small single portion of the rhizome can quickly develop into a new plant.

Fragrant Water Lilies benefit the ecosystem by providing shelter and habitat for fish and invertebrates. They also provide a cool and shady habitat for fish. Ducks and other waterfowl eat the seeds. One of the most important attributes of this native plant is the beauty they contribute to watershed.

There are two morphological types of *Nymphaea odorata lilies* (Palacio-Núñez, et al.) throughout the Media Luna Watershed, 1) "carpet shaped" these plants grow close together and form a dense covering over the silty or muddy bottom, and 2) "floating leaves" these plants are more dispersed with some leaves that reach the surface.

Submerged Plants

Submerged plants are rooted to the floor of the water system. Most of their leaves and stems are below the water surface although portions may lie on top of the water. Submerged plants have flexible stems. They collect small levels of sunlight because light energy is quickly reduced while traveling through the water column. Submerged plants can be seen clinging to the bottom of the canals where the swiftly running water causes the leaves to wave and somewhat dance along the canals floor.

Eel Grass
Vallisneria americana
Other names: celery grass, wild celery, water-celery, tape grass

Eel Grass is a typical submerged aquatic plant in the canals and channels flowing from Media Luna. Eel Grass leaves are about 2.5 cm one-inch-wide and can be up to 120 cm long. The leaves have curved tips and raised veins. The long ribbon-like leaves grow from stoloniferous clumps that are securely embedded underwater and spread by sending out runners. They are most abundant along cracks in the canals concrete. Individual fertilized white female flowers reach the water surface where the flower stalks curl into spirals and draw the flowers under the surface. The fruit is a banana-like capsule containing numerous small seeds. The seeds are a major source of food for turtles and other aquatic wildlife.

Sago Pondweed

Stuckenia pectinata

Other Common Names: broadleaf pondweed; duck grass; fennel pondweed; foxtail; Indian grass, old-fashioned bay grass; pond grass; potato moss; wild celery; fennel-leaved water milfoils; poker grass; orchard grass; string weed

Spanish - Potamogeton pectinatus

Sago Pondweed can grow up to 3 meters tall. It emerges from a slender rhizome and remains completely submerged except for the reproductive stalk that reaches above the water. Leaves are spiky, slender, filiform (threadlike) and resemble a pine needle. The leaves are 2-12 cm long and between 0.2 -1.5 cm wide. The flower is a spike with whorls 1 – 5 cm long. Fruits resemble beads and are small ovoid brown or yellowish bulbs, 0.3 - .5 mm long and look like a nut.

Sago Pondweed is considered a noxious weed in some recreational waters and irrigation ditches. It is also known to be a substrate that allows troublesome algae to attach. Dense growths of Sago Pondweed may limit flow and movement of fish as well as clog water intakes of mechanical irrigation units. Within the Media Luna Watershed, Sago Pondweed does not grow in dense patches and is not a problem. Frankly, it's hard to locate. Sago Pondweed does produce large crops of seeds and tubers that are consumed by waterfowl. The tubers and seeds are exceptionally nutritious, although leaves, stems, and roots are also consumed. It also provides juvenile fish and invertebrates cover from predators. Sago Pondweed also works as a nutrient buffer by using dissolved nitrogen and phosphorus for development. This function reduces algae blooms by making the nutrients unavailable for the algae.

Water Primrose

Ludwigia sp.

Spanish: duraznillo de agua

There are many species of Water Primrose (*Lubwigia*). Some are both submerged and floating plants. The species found around the Media Luna Watershed is a perennial plant with roundish leaves and reddish undersides. Other species of Water Primrose have lance-shaped leaves. The leaves are opposite. The fruits and yellow flowers are sessile without a peduncle. They have been observed flowering all year long within the watershed. They usually grow in shallow marshy areas and along the edges of lakes, canals, and channels as well as in ditches containing standing water. Water Primrose are generally removed by cutting within the watershed, but they are difficult to eliminate because they can become reestablished from seeds and remaining roots. Water Primrose can grow to become an aggressive invader of ponds and in other still-water conditions.

Emergent Plants

Emergent Plants grow along the shoreline and extend above the water surface. Cattails are an excellent example of an emergent plant. The stems of emergent plants are rigid or stiff. The most common species of emergent plants are hundreds of grasses, (Poaceae). Emergent plants are the ones that are most abused by visitors to Media Luna. If one examines the shoreline where swimmers enter and exit the water one will notice the bare dirt shorelines where emergent plants have been eliminated by human traffic.

Cattail

Typha sp.

Other common names: Reedmace, Catninetail, Punks, Corn Dog Grass

Cattails (*Typha spp.*) around the watershed are about 1.5 meters tall but can reach up to 2.5 meters tall in ideal conditions. They are one of the most common and easily identified plants within the Media Luna Watershed. Their leaves are flat to somewhat rounded and often have a slight twist to them. Cattails are well known for their dense dark brown spikes that emerge from the terminal end of the stem. The spikes are cigar-shaped and officially called catkins and can grow to a length of up to 30 cm. They flower in the spring and summer. The slimmer, upper part of the catkins comprises the male flowers while the lowermost parts are the female flowers. Cattails can grow while partially submerged in water or wet soil conditions. Cattails can spread through the production of seeds that are released from the catkins or by underground rhizomes. They have the capability to spread rapidly.

The submerged sections of cattails supply healthy environments for micro and macro invertebrates that in turn are used as food by fish and other wildlife species. They also provide protective cover and nesting areas for animals and birds. Cattails are sometimes thought of as a nuisance along the shorelines of the Media Luna Watershed. They perform essential functions in maintaining high-quality water by filtering runoff as it flows into the water system. They also help prevent shoreline erosion

GRASSES

There are many species of water grasses within the watershed. The greatest diversity of species is located along the south shore of Media Luna. The grasses, mixed with forbs, sedges and rushes are so thick that their roots form an underwater ceiling for cave-like features. A survey of these grasses was not possible because the region is blocked off as an environmentally sensitive area. Therefore, we will discuss the most commonly seen water grass, *Luziola fluitans,* often viewed along the lake shores, channels, and canals.

Water Grass
Luziola fluitans
Other Common Names: Silverleaf Grass, Southern Watergrass

Water Grass (*Luziola fluitans*) is a member of the "Poaceae" family. It is found along the shorelines of canals and channels as well as around the lake. It is a true perennial aquatic grass that resembles crabgrass. It is usually rooted in the shore and forms mats that creep out from the shore. Sometimes, the roots are submerged in water up to 1 meter deep. Water Grass prefers slow moving waters. Its leaf blades average 4 to 12 cm long and 0.6 cm wide. The top of the blade feels sandpapery to the touch. The stems are slender with the lower stems submerged and the upper stems floating. The light green leaf blades are flat and short. Water Grass blooms from summer to fall. It can tolerate droughts as long as the soil remains wet. It's tiny, infrequent, and whitish inflorescences are branched just above the leaves,

The submerged parts of Water Grass provide habitats for micro and macro invertebrates that fish consume as well as amphibians, reptiles, and birds. Birds also eat the seeds and leaves. Water Grass is controlled in the canals used for irrigation because it can clog up the water intake systems.

Common Reed

Phragmites sp.
Other common names: Reed Grass, Cane

Common Reed is a tall, robust grass that can grow to four meters in height. Common Reed grows in sizable clusters or colonies. The shiny gray-green leaves are arranged alternately along a stiff, smooth, vertical stem. At the leaf-sheath junction is long white hairs. The lance-shaped leaves are pointed, around 2.5 cm wide and 25 to 50 cm long. The flowers have brown to purple petals and appear from July through September. Sometimes the inflorescence is a silvery tan. The plume shaped flowers are surrounded by silky white hairs that are often draping to one side. The small brown fruit ripens in late July through October. Common Reed prefers wet habitats. They can be seen growing almost everywhere around Media Luna and its canals and channels. Thick rhizomes form colonies of Common Reed.

Common Reed can quickly grow into thick stands of stems that crowd out or shade other vegetation. Fortunately, the residents harvest reed and use it for fences, thatched roofs, and erosion control. This plant is also important to wildlife as it provides protection and cover. Many birds eat the seeds of Common Reed and nest within the colony.

Common Reeds, and the rhizomes. However, this plant is wilds protection and <u>cover</u>.

Figure 17 - Reed has many uses. It is used for fences and thatched structures throughout the entire watershed.

Giant Reed
Arundo donax
Other Common Names: Carrizo, Arundo, Spanish cane/reed,
Colorado River, Reed, Wild cane, and Giant cane
Spanish: caña común, caña de Castilla, cañabrava, carrizo

Giant Reed, *Arundo donax*, is an invasive grass that is native to the Old World. It usually grows to about 6 meters in height but in ideal conditions, it can exceed 10 meters. It resembles the Common

Reed, *Phragmites australis,* but can grow much taller. It has hollow stems 2 to 3 cm in diameter. The 30 to 60 cm long leaves are alternate and 2 to 6 cm broad with a tapered terminal end. Giant Reed flowers are upright with feathery plums 30 to 60 cm long and whitish when mature. Although it flowers, the seeds are usually not viable. They primarily reproduce by underground rhizomes. They prefer wetlands and can be seen growing along the canals and channels within the Media Luna Watershed.

The Giant Reed uses more water than native plants thus lowering groundwater tables. Giant Reed is raised as an ornamental grass. Recent studies have been done to promote it as a biofuel. It is used as building material. It is highly invasive and is responsible for decreasing plant biodiversity especially in riparian habitats. Left uncontrolled, it can transform healthy riparian plant communities into solid stands of Giant Reed. The Giant Reed is among one of the worst invasive alien species in the world.

Sedges
Cyperaceae (family)

Other Common Names: Deer Grass, Grassweed, Swamp Grass

There are many species of sedges. The number of species growing in the Media Luna Watershed is unknown. Sedges are monocots with over 5,500 species. They prefer damp soils and bear a resemblance to grasses. Sedges usually have upright or limp brown or green flower spikes located on the tips of their stems. A simple phrase used to help identify sedges is, "sedges have edges," referring to the fact that a good number of sedges have three-sided stems that are triangular in cross-section. Some of the larger sedges can reach up to 1.2 meters (4 feet) in height although most of the sedges that grow in the Media Luna Watershed are much shorter. Sedges are considered useful wildlife plants. Waterfowl and many small birds graze on the seeds of sedges.

Figure 18 - Note the triangular cross-section of most sedges.

TABLE 1

Guide for Differentiating Between Sedges, Grasses, and Rushes

Cyperaceae (Sedges)	Poaceae (Grasses)	Juncaceae (Rushes)
Stems usually 3-angled (or round, 4-angled, lenticular)	Stems round (terete)	Stems round (terete)
Stems usually with solid pith	Stems with solid nodes and hollow internodes	Stems with solid pith
Leaf sheaths closed	Leaf-sheaths open	Leaf-sheaths open
Leaves 3-ranked or spiral	Leaves 2-ranked	Leaves 2-ranked
Fruit an achene with bristles, bracts, may have tubercle	Fruit a grain with papery palea, lemma, and glumes	Fruit a capsule with tiny dust-like seeds

Spike Rush
Eleocharis sp.

Other Common Names: Creeping Spikerush

Rushes are members of the family "Juncaceae" a slow-growing, rhizome forming, herbaceous, flowering plant that somewhat resemble grasses and sedges. In general, keep in mind that "rushes are round, and sedges have edges" when attempting to distinguish between rushes and sedges. Spike Rushes are typically small (although some may reach up to a meter in height). Those found around the Media Luna Watershed are frequently 12 to 20 cm tall. They can grow completely underwater and may appear as a submerged plant. There are over 400 species of spike rush and are hard to identify without a detailed botanical key beyond the scope of this guide. Spike Rushes spread from rhizomes in shallow water and moist soils. A few rushes are annuals, but most are perennials. The florets on rushes are concentrated in terminal inflorescences with three sepals, three petals, 2 to 6 stamens and a pistil with three slim styles. The leaves are usually thin sheaths and may not be obvious during the initial examination. Developed fruits grow on the tips of the stem and are pointed with numerous seeds in a three portion chamber.

Wildlife and livestock consume spike rush and benefit from the reasonably high protein content. It also provides nesting cover for waterfowl. Ducks and other waterfowl eat the seeds. Spike rushes dense root mass provide soil stabilization along the shores of the lake and channels. The rhizomes form a matrix for many beneficial bacteria that help to maintain water quality.

Marsh Purslane
Ludwigia sp.

Other common names: marsh seedbox, water purslane, creeping
primrose, false loosestrife, Hampshire purslane, marsh ludwigia

Marsh Purslane grows in shallow areas along the canals, channels, and along the shores of Media Luna and adjoining wetlands. It is a native, dicot, semi-aquatic, perennial herb that likes moist or wet areas. Marsh Purslane is a species of flowering plant in the evening primrose family. The lance-shaped to oval leaves are oppositely arranged and green to red or purple in color. The plant grows from 7 to 40 cm in length and has a somewhat succulent appearance. When it is creeping out of the water, the leaves stretch across the ground. When it is in shallow water, it rises toward the surface where its taller stems and leaves frequently protrude above the water. When the stems reach above the water surface, buds develop from the axils of leaves. Each blossom has a short tubular calyx with four teeth, four short stamens, and a small style directly above the emerging ovary. Petals are usually absent. They produce small seeds in a capsule that contains many small seeds. The tiny seeds are thought to stick to the feathers or muddy feet of waterfowl that spread the seeds to new wetlands. This species primarily reproduces from asexual rooting plant fragments that are spread by currents.

Marsh Purslane is a common aquarium plant. It is also popular in water gardens and is even believed to have medicinal uses. It provides food for rodents and waterfowl. It also provides cover for amphibians and small fish. Because it is thought to obstruct water flow, Marsh Purslane is commonly removed from the canals and channels within the watershed. It is believed to be a weed in some parts of the world but not within the watershed. It is truly a beautiful plant that adds lovely colors in the areas it grows in, both in the water or on the shoreline.

Water Pennywort
Hydrocotyle spp.

Other Common Names: Indian Pennywort, Marsh Penny, Thick-
leaved Pennywort, Floating Pennywort

Water Pennywort (*Hydrocotyle spp.*) is a member of the Apiaceae family that includes carrots, celery, parsley, and many other plants that we consume. Water Pennywort is an aquatic/semi-aquatic relatively small perennial plant. It can be found throughout the Media Luna Watershed, particularly when the water current is slow. Water Pennywort will also grow in moist soils. The easiest way to identify Water Pennywort is by the dark green, lobed (3-7), round leaves that rise above the water's surface. The leaves are usually 2-6 cm in diameter. The lobe separations reach about mid-leaf. The leaf edges are relatively smooth with small depressions. The stems float to the surface where they sometimes form dense mats that may extend to the shoreline. The flowers are small, white, greenish, or sometimes yellow that arise from the leaf base in clusters of 5-10 at the tops of 1-5 cm long flower stalks. Each flower has five tiny petals. The fruits that carry the seed are small achenes that can float making the dispersal of the seeds more widespread. The root can be floating or rooted in the substrate.

As with all aquatic plants, the submerged portions provide habitats for many micro and macro invertebrates that provided food for fish and other wildlife.

Photos by Lebendkulturen de and Pi-Lens/Shutterstock.com

Algae

Algae are abundant in the watershed. Because they are microscopic, many are not even noticed by the unaided eye. When some planktonic algae numbers are high enough, they can make the water appear green due to the chlorophyll. They can also form clusters of red or white slime-like substance that settles on the bottom of the lake. Small blooms have been identified by scuba divers since the formation of the park. The recurring incidence of visible algal blooms frequently indicates that lake nutrient levels, especially phosphorus, are too high. Filamentous algae also thrive throughout the watershed and can be easily identified with a basic microscope. Chara (stoneworts, musk grass, or skunkweed) is another type of algae that can be found growing throughout the watershed. Chara resembles vascular plants, but Chara is a green alga that lack true leaves or roots.

Chara is an advanced form of algae often mistaken for higher vascular plants.

Algae are an excellent source of food for fish and other organisms found in the watershed. It is important to keep a keen eye out for various algae growths because some species of algae (blue-green algae) release toxins. In recent years, scuba divers have noticed an increase in algae growth in Media Luna. These unsightly scums are either caused by tangled masses of filamentous green algae or by blooms of unknown blue-green algae that cover portions of the lakes bottom. These visible occurrences of algal blooms may indicate nutrient levels, especially phosphorus, are too high. Blooms could be caused by overexploitation of the lake and should be monitored by biologists.

Planktonic algae include of diatoms, blue-green algae, and green algae. They are microscopic in size and are the basis of the aquatic food chain in the watershed. With higher levels of light, phosphorus, and nitrogen come an abundance of planktonic algae. Although, an abundance of nutrients can result in low water clarity thus limiting the growth of algae. Also, low ratios of nitrogen to phosphorus will favor the growth of blue-green algae over the most edible forms of green algae.

Colonies of filaments form filamentous algae. The algae seen growing on the surface of the water are usually filamentous algae, although some filamentous algae become submerged in the shallow water attached to the bottom or other structures such as rocks. They provide habitats for numerous micro and macro invertebrates. Invertebrates are consumed by many fish, reptiles, birds, and amphibians. High levels of nutrients can cause filamentous algae to grow in abundance resulting in low levels of dissolved oxygen due to the dying algae that consume oxygen from the water. Low levels of dissolved oxygen also result in a reduction of invertebrates. In the Media Luna Watershed, these excessive nutrients may come from runoff of animal manure, fertilizers or other chemicals used in crop fields, or from septic systems. An overabundance of algae stresses fish and other critters living in the water system. When nutrients get out of control, an algae bloom may occur.

Figure 19 - Small floating algae blooms observed in recent years. Photo by Ossiel Martinez.

Chapter 6

FISHES OF THE MEDIA LUNA WATERSHED

Only when the last tree has died, and the last river has been poisoned, and
the last fish been caught; will we realize we cannot eat money.
Cree Indian Proverb

The fishes listed in this book are based on Dr. Palacio–Núñez et al. findings while performing surveys to determine the value of diurnal fish and birds as bioindicators. Also included is the Rio Verde Catfish, *Ictalurus mexicanus*. This nocturnal species is included because it is often observed during night-time scuba dives made by the author. The list is all-inclusive of species known in the Media Luna Watershed at the time of printing. The photos were taken by Seth Patterson unless otherwise noted.

According to IUCN, International Union for Conservation of Nature, some of the fishes in the Media Luna Watershed are threatened to critically endangered. The IUCN maintains a Red List that assesses the conservation status of species, subspecies, varieties, and even selected subpopulations on a global scale. It identifies threatened species and serves as a barometer of life on earth with the goal to promote their conservation. The IUCN Red List is a useful instrument to inform concerned groups/organizations and to support an action for biodiversity preservation and policy change, critical to protecting natural resources. Fishes of the Media Luna Watershed are included in the Red List although several species have not been evaluated.

Not included as a resident of the Media Luna Watershed is the Flatjaw Minnow, *Dionda mandibularis*. This species existed about 8 km southeast of Rio Verde in a channel that extends from a cold-water spring called "Charco Azul" and still has not been found within the Media Luna Watershed. The Flatjaw Minnow is listed as critically endangered. A search for the Flatjaw Minnow would be a significant research project and would settle the question of its existence in the Media Luna Watershed.

See Table 2 for a list of fishes that are endangered based on the Mexican Official Standard – 059 (INE 2002). Note that five species have not yet been evaluated. Those species not yet evaluated offer an opportunity for researchers but also demonstrates the need for further research of the Media Luna Watershed.

TABLE 2

Status of Fish Species Present in the Media Luna Watershed, Their Distribution and Extinction Risk Category According to the Mexican Official Standard - 059 (INE 2002)

FAMILY	SPECIES	DISTRIBUTION	RISK
Characidae	*Astyanax mexicanus* Filippi	3	NL
Cichlidae	*Cichlasoma bartoni* Bean	1	E
	C. cyanoguttatum Bird and Girard	4	NL
	C. labridens Pellegrin	2	E
	Oreochromis sp	5	NL
Cyprinodontidae	*Cualac tessellatus* Miller	1	E
Cyprinidae	*Tampichthys dichromus/ Dionda dichroma* Hubbs and Miller	1	T
Goodeidae	*Ataeniobius toweri* Meek	1	E
Poecilidae	*Gambusia panuco* Hubbs	4	NL
	Poecilia latipunctata Meek	4	CE
	P. mexicana Steindachner	4	NL
Ictaluridae	*Ictalurus mexicanus*	2	T

Table Guide: **D** = distribution: **1** = Endemic to the Valley of Rio Verde, **2** = Endemic to the region, **3** = Native with wide natural distribution, **4** = Native to NE of Mexico, introduced to Media Luna, **5** = Native of another country. Based on the Mexican Fishes List

(Espinosa-Pérez *et al.* 1993). **R** = Risk: **NL** = not listed, **T** = threatened, **E** = endangered, **CE** = critically endangered. (Ref. Palacio-Núñez, Jorge, 2010) * IUCN Red List

Photo by Seth Patterson

Mexican Tetra
Astyanax mexicanus

Note: Not the Blind Mexican Cave Fish or Mexican Cave Tetra
Spanish: Sardina Ciega

The Mexican Tetra is a small (up to 100 mm) minnow-like silvery fish with a blackened lateral stripe that is bigger near the caudal base and narrows at the caudal fin. They also have a small dorsal adipose fin between the dorsal fin and the forked tail. During the breeding season (usually late spring and early summer) the male's dorsal fins and tail become yellowish to orange-red, and small hook-like features may appear on the anal fins.

It should be noted that breeding seasons can vary in Media Luna when compared to other locations. Most of the species that change color during breeding seasons can be seen displaying breeding colors during the entire year. Perhaps, this is due to the constant warm water temperatures found in the watershed.

Mexican Tetras are a schooling species with a wide variety of aquatic habitats including springs, streams, lakes, reservoirs, irrigation ditches, and small pools. In Northeastern Mexico, the species are omnivorous with plants, algae, aquatic insects, and occasionally fish, comprising the central part of their diet. Mexican Tetras are an opportunistic invasive species that can displace other species in ecologically stressed environments. They prefer warm water that is less than three meters deep.

Scuba divers and snorkelers can readily recognize this fish by the shiny silver flash they produce and their colorful tails as they quickly swim away. They are abundant underneath the docks at Media Luna

Photo by Ammit Jack/Shutterstock.com

Tilapia
Oreochromis sp

Other Common Names: Mudfish, Kurper mojarra, mojarra Africana
Spanish: Tilapia

Tilapia is the generic name of a group of cichlids endemic to Africa. Tilapias are an exotic species, introduced into Media Luna Lake about 20 years ago. Tilapias look a lot like bream, sunfish or crappie. They usually have a two-part lateral line with the front part higher on the body than the rear part. They have long, pointed, unbroken dorsal and anal fins. There are frequently wide vertical bars along the sides of the young and sometimes adults. Both males and females can grow up to a length of 40 cm. They can develop a broad range of skin colors due to the environment, stage of sexual maturity, and food source.

Tilapias are nest builders, and the parents guard fertilized nests. Some species are mouth brooders where after fertilization the parents pick up the eggs in their mouth and hold them for incubation and several days after hatching. They are aggressive nest builders and can reduce the population of other fish nest builders due to competition for nesting sites. Most of the Tilapia in Media Luna have a row of black dots along the lateral lines.

Tilapias have the capability to increase their population quickly in a favorable habitat, such as Media Luna. They have an elevated tolerance to low dissolved oxygen conditions and can tolerate a broad range of temperatures (6.5 – 42.5 degrees C). Their most favorable temperature range is 20 to 30 degrees C. They eat plankton, aquatic plants, and immature stages of many flies, beetles (adults and immature), mayflies, caddisflies, stoneflies, dragonflies, aquatic worms, snails, leeches, larval fish, and decomposing organic matter. Once they have depleted a food resource at one location, they can quickly move to another. Because of their feeding habits, they can have negative impacts on aquatic plant communities as well as other fish species.

Tilapias have become the most common fish seen by scuba divers over the past few years. They are extremely invasive and can take over a small lake such as Media Luna if conditions are right. Tilapias are remarkably adaptable to many environments, and this makes them particularly prone to becoming an environmentally taxing invasive species. The can quickly propagate and cause problems for native flora and fauna by upsetting the ecological equilibrium. As seen in Media Luna, they can cause a buildup of turbidity, a result of their fondness of digging. Turbidity can reduce the accessible light in the lake that affects organisms relying on photosynthesis.

Photo by Gualtiero Boffi/Shutterstock.com

Panuco Gambusia

Gambusia panuco

Other Common Names: Rio Panuco Mosquitofish, Panuco
Mosquitofish, Panuco-Kärpfling, Panuco Basin gambusia
Spanish: Guayacón del Pánuco

The Mosquitofish was introduced into the Media Luna Watershed. They are a small dull gray to brown freshwater fish with a large abdomen. The males may reach a length of 4 cm and the females seven cm. They have curved dorsal and caudal fins and an inverted mouth that lower jaw points to the surface. The anal fins on females are like the dorsal fins, but the anal fins of males are pointed and used to deposit sperm inside a female. They do not have a forked tail.

They are an introduced species in Media Luna that prefers shallow flowing and standing waters. They live throughout the entire Media Luna Watershed. Spawning takes place in spring. The female can store sperm from multiple males. Larval and adult insects and zooplanktons are their primary food.

Some believe the mosquitofish was introduced as a biocontrol to help reduce the mosquito populations. Their introduction has an adverse effect on native species that consume the same foods. One characteristic of the mosquitofish is their ability to survive harsh environments such as low dissolved oxygen levels, warm water, and high salt concentrations.

Gambusia. Panuco and Gambusia regani are considered the same form of species by some authors.

Illustration by Israel Ivan Garcia Martinez.

Bicolor Minnow

Tampichthys dichromus, (Dionda dichroma)

Spanish – Sardinita de Rio Verde, Carpa bicolor

The Bicolor Minnow is a member of the Cyprinidae family, the family of carp and minnows, also known as cyprinids. The Bicolor Minnow is a small ray-finned silvery fish found in shallow water, usually not more than a meter deep. They lack teeth in the mouth; instead, they have teeth in the throat for grinding up the food against a bone situated at the back of the cranium. The Bicolor Minnow has one dorsal fin with less than ten rays and a pair of pelvic fins. They have a distinct lateral line. Their color is rather sooty above the lateral line and noticeably lighter below. The contrast of colors gives it a bicolor appearance. In some males, the fins appear to have a bluish edge.

The Bicolor Minnow can grow to a length of 3 to 4 cm. They usually spawn in crevices on rocky bottoms during the spring. They are omnivores that primarily eat algae and other small organisms found in slow moving waters. They mainly live in the channels and canals.

The Bicolor Minnow at Media Luna is a threatened species. The Bicolor Minnow is a critical species in the aquatic community because they are a food source for larger fishes as well as birds (especially wading birds and kingfishers).

Photo by Seth Patterson

Checkered Pupfish
Cualac tessellatus

Other Common Names: Cachorrito de Media Luna, Media Luna Pupfish
Spanish: Cachorrito de Media Luna

The Checkered Pupfish is an endemic and endangered species found primarily in Media Luna's lake and natural channels. It only lives in Media Luna Springs in the Rio Verde Valley, San Luis Potosi. It is silver/brown and usually has small black spots down its sides. Males have dark spots along the flanks and may be more random (thus its name "tessellated' meaning mottled or checkered appearance). Males also have orange to yellow pelvic/anal fins and a bluish dorsal fin that is peppered with beautiful dark spots. The male has a dark caudal fin that intensifies in color during spawning season. Females are much plainer, missing excessive flank markings and showing much dimmer pigmentation compared to the male.

Checkered Pupfish may grow up to eight cm, but the average size of an adult is around five cm. Eggs are attached to algae or other plant surfaces using a small filament. They lay in areas where plenty of cover is present for the young. Checkered Pupfishes are micro predators feeding on tiny aquatic crustaceans, worms, insect larvae zooplankton, and other plant material. They are not usually found in large numbers and inhabit shallow waters along the lake's edge and in the channels.

Photo by Kazakov Maksim/Shutterstock.com

Rio Grande Cichlid

Herichthys cyanoguttatus (Cichlasoma cyanoguttatum)

Other Common Mames; Rio Grande Perch, Texas cichlid, mojarra de
Norte, Perlscale Cichlid, Green Texas Cichlid, Blue Texas Cichlid
Spanish: mojarra del Norte

The Rio Grande Cichlid, anies is identified by its presence of blue or blue-green iridescent to whitish spots all over its body giving them a speckled look. It also has a broken lateral line; lateral-line separates near the end of the caudal fin and starts again posterior to its origin. Occasionally, it may have one to four black spots along its lateral line extending to the base of the tail. It is an oval-shaped cichlid with a slightly concave forehead and an arched back. The dorsal and anal fins are long and tapered extending beyond the fleshy section of the tail. Adult males are usually larger than females and may develop a pronounced hump on the head. Adults typically range up to 20 – 30 cm in length.

The territorial and aggressive Rio Grande Cichlid is considerably pollution tolerant. It eats both plant matter and animals (omnivore) such as small fish, worms, insects, and crustaceans. Adults are known to eat large quantities of other fish's eggs. They prefer warm water, (20°C - 33°C) especially spring-fed warm water like Media Luna.

Rio Grande Cichlids are an egg-laying species and spawn in the early spring. Both parents protect their fry (young fish). Rocks are the preferred substrate for spawning and may be a limiting factor in Media Luna. Breeding pairs will attack other fishes in the nest's surrounding area but tend to be more aggressive against other cichlids.

Photo by Seth Patterson

Barton's Cichlid
(Herichthys bartoni - Cichlasoma bartoni

Other Common Names: Skunkfish, Media Luna Cichlid
Spanish: Mojarra Caracolera

The Barton's Cichlid has a relatively narrow body with a curved head at the front. Males have a small hump on the front of the head. Their snout is long and somewhat pointed with thick lips and mainly anterior teeth. The pelvic fins are short, and the dorsal and anal fins slightly rounded. While in normal coloration, the Barton's Cichlid is gray-brown with an inclination of a lackluster yellow with a row of black blotches running from behind the eye to the base of the caudal fin. Some of the scales on the lower half of the body may have a blue tint to them. Spawning may occur any time of the year. During breeding, their color changes to a stunning bicolor black and white in the males; black on the bottom half and bright white on the top half, (thus its nickname, skunk fish).

Barton's Cichlids inhabit clear thermal waters with a sandy, silty, or muddy substrate (bottom). They are egg laying species that prefers to spawn along the banks of canals in caves or hollows and around rocks. They usually produce a high number of offspring (up to 100) where the adults guard the fry with the female near the nest and the male outside the nest.

The males grow to a length of up to 18 cm while the females are considerably smaller, growing to a length of 12 cm. They are omnivores with a diet that consists of smaller fish, aquatic insects, algae, decaying leaves, detritus, snails, and various other invertebrates.

Barton's Cichlids are endemic to the upper Rio Verde Media Luna system and listed as a threatened species. They are a favored aquarium fish and have suffered from captures of people wanting them for their aquariums. Some springs where they naturally existed have dried up due to droughts.

Photo by Seth Patterson

Bluetail Goodea
(Ataeniobius toweri)

Other Common Names: Mexcalpique cola azul, Bluetail
Splitfin, Striped goodeid, Mexican goodeid
Spanish: Mexcalpique cola azul

The Bluetail Goodea is occasionally a light gray to slightly silver-grey or sometimes with brownish body parts above and lighter below the lateral line. They are speckled along the sides where the dark and light colors meet. Some can show a random number of vertical bars on the posterior half of their body extending from the last part of the belly to the caudal fin. Some males may have blue fins all year long. Although, during courtship and breeding periods the males develop a metallic blue color that gets brighter at the tail; hence the name Bluetail. Females remain more mildly brownish-grey colored and at times have two horizontal stripes along her sides.

Bluetail Goodea prefers water with slight or no currents. They can be found along the shores of the lake and channels and in irrigation ditches. They prefer clear water, especially around water lilies. They are omnivores that eat small crustaceans and filamentous green algae. Females grow to 10 cm while males only grow to a length of 7 cm. Six to twenty fry are born alive (livebearers) after two months of gestation.

The Bluetail Goodea is classified as critically endangered and facing an elevated threat of becoming extinct; mainly due to a loss of habitat. Scuba divers seldom see them at Media Luna.

Photo by Seth Patterson

Herichthys Labridens complex
Chichlasoma Labridens

Other common name: Chichlasoma Labridens
Spanish: Mojarra huasteca

Herichthys Labridens complex is a group of subspecies with a shape similar to the Barton's Cichlid. The Riverside (yellow labridens) that are found mainly in Rio Verde River have an elongate body and rounded anterior profile. The morph phase found only in the Media Luna Watershed possesses a steeper forehead and longer snout. They have a small to medium sized mouth that is slightly angled downwards and the upper jaw projects over the lower one. They often appear to be sprinkled with shiny blue areas. The breeding colors of the *Herichthys Labridens* are a magnificent bright yellow on the dorsal portion with silky black coloration on the ventral parts.

Herichthys Labridens can grow up to 20 cm where the female is usually smaller than the male. They feed on snails located on the substrate. When the meat of the snail is crushed out, the shell debris is ejected out of the mouth and gills. They are known to spawn throughout the year and will build a nest under leafy vegetation or on a rock. Both parents are very protective of the nest and fry and will attack approaching predators, including scuba divers.

Photo by Seth Patterson

Shortfin Molly
(Poecilia Mexicana)
Other Common Names: Mexican molly, Atlantic molly, orange fin molly, Campeche, Tabai

The Shortfin Molly is an introduced species to Media Luna although it is native to Northeast Mexico. Perhaps it got into the Media Luna Watershed through the aquarium trade. The laterally compacted Shortfin Molly has a thick body with a gentle yet fixed slope down from the anterior tip of the dorsal fin to the snout. They may show deepness in the trunk, especially in females as they swell when carrying young. They have noticeable spots and bands arranged along the body and fins. Shortfin Mollies have broad caudal and dorsal fins. The common name for this fish is based on the fact that males do not have extended dorsal fins. They are a small fish but robust in appearance and strong swimmers. Males are more colorful and may have a yellow edge on the tail.

Shortfin Mollies like warm springs, canals, and weedy ditches as well as stream pools. They will eat detritus, algae, diatoms, and plant matter. Females go through internal fertilization and live birth. They can grow to 20 – 42 mm but the growth is significantly dependent on the quality of their habitat.

Photo by Seth Patterson\

Broadspotted Molly
(Poecilia latipunctata)
Other Common Names: Molly del Tamesi, Topote del Tamesí

The Broadspotted Molly is another introduced species to Media Luna Watershed. It is another species that is native to Northeast Mexico. The Broadspotted Molly is an odd molly species because it has features of both the Shortfin and Sailfin mollies. Officially, it is classified as a Shortfin species, yet the male courtship behavior is like that of a Sailfin. Their body structure is the same as the Shortfin Molly, somewhat robust and deepness in the trunk. The males are silvery in the ventral portion and greenish on the dorsal areas. The females are a bit more mundane in color. They are distinguished from the Shortfin Molly by a row of dark spots that travels along the lateral line.

Broadspotted mollies prefer warm, calm, and vegetated waters. They are live-bearers and can grow to lengths of 25 – 45 mm. Broadspotted mollies are endangered.

Rio Verde Catfish
Ictalurus mexicanus
Spanish: Bagre de Rio Verde

The Rio Verde Catfish, an endemic to central Mexico, is listed by the International Union for the Conservation of Nature (IUCN) as a threatened species and vulnerable. They can grow up to 26 cm in length. They have pronounced barbels that bear a resemblance to the whiskers of a cat. The long barbels contain sensors for smelling and for sensing dangers that may exist in their surroundings. They are mainly bottom feeders where they eat smaller fish, aquatic insects, crayfish, and frogs. The Rio Verde Catfish does not have scales, but they do have a tough mucous covering. The mucous coating is blackest on the dorsal areas and lighter on the ventral areas. Care must be taken when handling these fish because they have sharp spines on their dorsal fins and the sides of their pectoral fins. Both males and females favor rocky or sandy bottoms in calm to medium flowing waters. In the Media Luna Watershed, they are usually observed around sunken logs or other debris laying on the or around the opening of the springs. Females typically lay 10 to 90 eggs near the surface where they are safe from other bottom-dwelling aquatic creatures. The Rio Verde Catfish is seldom seen by snorkelers or divers because they usually stay hidden during the day and only come out at night to feed.

Chapter 7

BIRDS OF THE MEDIA LUNA WATERSHED

There is nothing in which the birds differ more from man than the way in which they can build and yet leave a landscape as it was before. Robert Lynd, *The Blue Lion, and Other Essays*

Mexico is blessed with some of the best birding sites in the world. There are more than 550 species of birds in the state of San Luis Potosi. The area covered in this chapter includes the entire Media Luna Watershed. The listed species are those seen throughout most of the year and are dependent on the watershed for food, breeding, and shelter. They consist of waterfowl and one raptor. The 19 bird species discussed in this book were chosen based on surveys conducted during the time this book was being written and surveys performed by Dr.Jorge Palacio–Núñez et al. in his study of birds as bioindicators. Not described are neotropical migrating birds and residents of the area that are not dependent on the watershed. We recommend the following websites for a comprehensive list of all birds one might see in the Media Luna Watershed:

http://www.birdlist.org/nam/mexico/san_luis_potosi/san_luis_potosi.htm - Birds of SAN LUIS POTOSI, Pájaros / aves de SAN LUIS POTOSI. The information from BIRDLIST is based on research from literature, the official national checklists of the birds, information from birdwatchers and extrapolation of our data.

http://avibase.bsc-eoc.org/checklist.jsp?region=mxsl&list=clements - This list includes bird species found in San Luis Potosí and is based on the best information available at this time. It builds on a wide variety of sources collated over many years by Denis Lepage and hosted by Bird Studies Canada, which is a co-partner of Birdlife International.

Great Egret
Casmerodius albus
Other Common Names: Great White Egret, Common Egret, Large Egret or Great White Heron
Spanish: Garceta Grande

The Great Egret is a large white heron found wading in the water anywhere within the watershed. They stand up to 106 cm tall. They have a yellow bill and black legs and feet although the bill may become darker, and the legs lighter during the breeding season. Males and females appear the same in appearance while juveniles look like non-breeding adults. The Great Egret is usually found wading in shallow water where it eats fish, frogs, small mammals, small reptiles, and insects. They use their bill as a spear their prey and are exciting to watch as they stand motionless or slowly hunt for their victims.

Right Photo by David Osborn/Shutterstock.com

Black-Crowned Night Heron

Nycticorax nycticorax
Other Common Names: Night Heron
Spanish: Martinete Común

The Black-Crowned Night Heron has red eyes, short yellow legs, a black crown, and black back. The remainder of the body is white or gray with pale gray wings and white underparts. They have two or three white plumes (lores) extending from the back of their head and erected in salutation and courtship displays. The males are usually a bit larger, but both sexes are similar in appearance. They are not the long, slim, birds one often envisions when thinking of herons; instead, they are stocky and short birds with short bills, and legs. Juveniles are dull grey-brown with pale spots. Young birds have orange eyes and duller yellow-green legs. Black-crowned Night Herons get their name from their habit of standing on the water's edge waiting to capture some prey during nighttime hours or early morning. They eat small fish, crustaceans, frogs, insects, small mammals, and small birds.

Great Blue Heron

Ardea herodias
Other Common Names: Great White Heron (white morph),
Wurdemann's Heron, Blue Crane, Blue Gaulin
Spanish: Garza Azulada

There are two morphs found in Mexico, the dark morph, Great Blue Heron, and the white morph that is sometimes called the Great White Heron. They are located near the lake, along the canals, and in a marshy area. They can grow to a height of 115 cm's making them the largest heron in North America. They are bluish-gray above with a whitish head and slender black plumes (lores) behind the eye. Their bill is long, thick, yellow-gray and sometimes orange below and darker above. The ventral parts are pale with white undertail coverts and gray-black primary and secondary feathers; although this feature can usually only be seen in flight. Juveniles are more mottled and have less defined head markings. They eat fish, amphibians, reptiles, and insects.

Green Heron

Butorides virescens
Spanish: Garcita verdosau

Other birders have identified the small Green Heron that can be seen around the watershed as the striated heron (*Butoride striata*) although the authors did not identify this species. We did identify

the Green Heron, (*Butorides virescens*), a species very similar in appearance to the striated heron. Therefore, the Green Heron will be described in this chapter. Collectively, these two subspecies are sometimes referred to as "green-backed herons."

The Green Heron (*Butorides virescens*) is the smallest heron found within the watershed.

Cattle Egret – exotic
Bubulcus ibis

Other Common Names: Buff-Backed Heron, Elephant Bird, Rhinoceros Egret, Hippopotamus Egret

Spanish: Garcilla Bueyera

The Cattle Egret is not confined to the wet areas of the watershed but can also be seen in fields of grazing cattle. They are a medium-sized (76 – 86 cm) white bird with small plumes (lores) on

Photo by Bonnie Taylor Baily/Shutterstock.com

their heads, chest and backs during the breeding season. They have a somewhat short thick neck and a short, stout, yellow to red-orange bill. Sexes appear the same although the males are a bit larger with slightly longer breeding plums. Juveniles lack the plumes and have a black bill. They nest in colonies and are often seen roosting in the Media Luna region. They feed in grasslands, farmlands, and wetlands. The Cattle Egret removes ticks and flies from cattle and eats them. This particular relationship benefits both species.

Photos by Gerald Marella and David Osborn/Shutterstock.com

Snowy Egret
Egretta thula

Other Common Names: Garceta Nívea
Spanish: Garceta Nívea

Snowy Egrets are a small to medium sized white bird, 60 to 70 cm tall that prefers lakes, ponds, streams, and canals. The males are slightly larger than the females, and both sexes have yellow eyes. They have a black bill and bright yellow buffs (lores). Snowy Egret's black legs are long and slender with yellow feet. Once they reach breeding season, they develop long, delicate plumes on their breasts, and their feet change from yellow to orange. They breed in the spring when the male can be seen performing courtship displays and vocalizations to attract females. The males display is characterized by pumping his body up and down with his bill pointed straight up towards the sky. They are often seen hunting for prey at dawn and dusk with other wading species of birds. Compared to other wading birds, the Snowy Egret has one of the widest range of food selections. The choices include worms, both aquatic and terrestrial insects, crabs, shrimp, crayfish, snails, fishes, frogs, snakes, lizards, and toads.

There are four distinguishing characteristics that differentiate the snowy egret from the cattle egret: 1) the snowy egret is larger, 2) during the breeding season the snowy egret has red lores, 3) the snowy egret has black bill and legs, but yellow feet and, 4) the snowy egret has a slender bill.

Tricolored Heron
Egretta tricolor

Other Common Names: Louisiana Heron
Spanish: Garceta Tricolor

Tricolored Herons were once called the Louisiana Heron, so some guidebooks may yet list it as the Louisiana Heron. Tricolored Herons prefer shallow waters and are often seen feeding with other wading birds. Tricolored Herons are beautiful small to a medium sized bird that stands 66 – 76 cm tall. Their legs are pink during the breeding season and yellow during non-breeding seasons. Their black tipped pointed bill is typically yellow but turns blue during the season of reproduction. Adults have a blue-gray head, neck, and back. Sometimes, a chestnut color can be seen along the neck. They have a white line along their necks and a white belly. During the breeding season, they have long blue plumes on the head and neck and buff ones on the back. Their legs are greenish or yellow in color.

Juveniles are more colored than adults and exhibit a rich chestnut head and neck with more chestnut coloring mixed in with the blue-gray back. Tricolored Heron eats fish, crustaceans, reptiles, and insects.

Photo by jo Crebbin/Shutterstock.com

Yellow-Crowned Night Heron
Nyctanassa violacea

Other Common Names: American Night Heron, Squawk
Spanish: Martinete Coronado

Yellow-Crowned Night Herons are a stocky small-medium sized heron that stands 40 – 50 cm high. They are a threatened species and can be found in all types of wetland habitats, flowing or standing water. They have gray bodies, black faces and bill with white cheek patches and a yellow crown. Both sexes appear similar during all seasons. Yellow-Crowned Night Herons have short yellow legs and red eyes. The juveniles are brown with small white spots on their wings and thin steaks on their ventral portions. They hunt from sunset to sunrise for crustaceans but also consume mollusks, insects, frogs, and fish.

Wood Stork
Mycteria americana

Other Common Names: Wood Ibis
Spanish Name: Cigüeña americana

Wood Storks inhabit the marshes and streams of the Media Luna Watershed and are a threatened species. Adult Wood Storks measure over one meter tall. They have a black bill and scaly looking featherless head and neck. Their bald and black face and downward tipped thick bill are their most distinguishing characteristics. When flying, their neck sticks straight out compared to the "S" shape of a heron. Wood Storks are mostly white except for some primary, secondary, and tail feathers that are black. They have gray to dark gray legs and pink feet. The Juveniles have a pale yellow bill and a dull gray head and neck. Wood Storks eat fish, mollusks, snails, frogs, and insects. Wood Storks often gather in groups of 12 -20, roosting in trees along the east channel.

Photo by Sam Aronov/Shutterstock.com

Neotropic Cormorant
Phalacrocorax brasilianus

Other Common Names: Olivaceous Cormorant
Spanish: Cormorán Biguá

Neotropic Cormorants are often seen roosting in trees near Media Luna. They live in many different habitats but prefer ones near deep water. Neotropic Cormorants need perches for sunning and drying their wings. They are a dark black color and stand 62 - 68 cm tall. They have a pale green throat patch at the base of the bill with a white "V" outline. Sometimes, a green to purplish luster can be seen on the upper wings and backs of adults, juveniles do not show the luster and may show some brown plumage. Both sexes appear the same. They have a dark, hooked bill that is yellowish at its base. They dive for their food in shallow and deep water where they find dragonfly nymphs, fish, frogs, tadpoles, and crustaceans.

Photos by Jan Nor Photography and Dave Montreuil /Shutterstock.com

Anhinga
Anhinga anhinga

Other Common Names: Snakebird, Darter, American Darter, Water Turkey
Spanish: Anhinga Americana

Anhingas may be seen anywhere near water habitats within the Media Luna Watershed. They prefer areas that include shrub and tree-covered islands and shores; especially near deep water. They can grow up to 85 cm tall. Their head appears small when compared to the long snake-like neck. Its sharp, serrated bill is dull yellow and pointed. They have a body shape of a Cormorant with black and greenish upper surface wings showing a long white patch. The tip of the tail has white feathers, and the back of the head and neck has grayish, white, and purple feathers. The upper back and wings are usually streaked with white. Anhingas have short legs and web feet and look like they are crawling when on land. They are often seen resting in bushes and trees over water holding their wings out to dry. They do not have oil glands for waterproofing, so when they swim, their entire body seems to be underwater with just their long slender neck and head showing. Juveniles show vast areas of brown. Anhingas primarily eat fish but are also known to consume crustaceans and insects. They spear fish amid their flanks with a rapid plunge of a partly opened bill.

The male starts the building of a nest in the fork of a tree before he has located a mate. Once he finds a mate, the female finishes building the nest. The nest will develop a layer of white excrement over time. Although, they are often seen soaring elegantly high in the sky. Most people confuse them with vultures but with a good pair of binoculars, one can see the long neck and profile of an Anhinga. Cormorants are sometimes confused with Anhingas. Cormorants will have shorter tails, shorter bills, and will lack the bright wing patches.

Black-Bellied Whistling Duck
Dendrocygna autumnalis

Other Common Names: Black-Bellied Tree Duck
Spanish: Suirirí Piquirrojo

Black-Bellied Whistling Ducks were once called the Black-Bellied Tree Duck. They prefer shallow freshwater lakes, marshes, ponds, and canals. They can also be seen on cultivated land with plentiful vegetation where they feed mainly on the seeds. The most probable location to spot a Black-Bellied Whistling Duck is along tree-lined bodies of water like the irrigation canals of the Media Luna Watershed. Yes, this duck perches in trees. Their length ranges from 47 to 56 cm. They are a medium-sized duck with a long neck, long pink legs, and a gray face. Male and female Black-Bellied Whistling Ducks are similar in size and color. Both have a black belly and chestnut nape and chest. Many have a chestnut cap on the top of their heads. They have a white eye-ring and a bright orange bill. Many have a large white patch on the back of their wings.

One can often hear them in flight whistling "pe-che-che" as they fly overhead. Ducklings stay with the parents for around eight weeks. Juveniles are like adults but duller and grayer with a gray bill and feet. They eat plant material, insects, and some aquatic invertebrates. They will sometimes feed on submerged vegetation in shallow waters. They like recently harvested fields where they eat leftover seeds and disturbed insects. They usually don't build nests as they lay their eggs on naturally deposited debris. They will nest in cavities in trees and on the ground. Often, females will lay their eggs in other Black-Bellied Whistling Duck's nests, a behavior called, "egg dumping." Major predators of the ducklings in the watershed are raccoons, snakes, owls, and fire ants.

The Black-Bellied Whistling Duck is often confused with being a goose. It has a gooselike long neck and long legs. These features make it easy to separate from other ducks.

Photos by Larsek and Agustin Esmoris/Shutterstock.com

Pied-Billed Grebe
Podilymbus podiceps

Other Common Names: None found
Spanish: Zampullín Picogrueso

The Pied-Billed Grebe resides in all areas of the Media Luna Watershed. They prefer habitat with emergent or aquatic vegetation. They are small diving water bird with a brown head and body and a whitish tufted rump. They have a small head and chicken-like short but thick bill. The bill is somewhat pale but has a black ring around it in the summer. Pied-Billed Grebes have a black patch on their throat in the summer with a whitish outline and a white ring around the eye. Their winter plumage tends to be pale on the throat with a fleshy colored bill with no black ring. The black ring on the bill is associated with the breeding season (summer). The juveniles are similar to a winter adult with dark and pale stripes on the face and a dull orange/fleshy bill. They lay 5-7 white eggs that appear stained brown in a concealed floating mass of old marsh vegetation. Adults average around 35 cm long. They eat small fish, crustaceans, aquatic insects and their larvae. Scuba divers see Pied-Billed Grebes diving underwater near the shores of Media Luna. When disturbed, the Pied-Billed Grebe will dive into the water rather than fly away, a practice that has earned it the label "Hell-diver."

Pied-Billed Grebes are common along the shores of Media Luna. They live amongst the thick aquatic vegetation that grows out of the water along the channels and canals. Pied-Billed Grebes are more abundant in Media Luna during the winter months because of less human activity and transients stopping off during migrations. There are also resident populations within the watershed. Their voice is a series of cuckoo-like notes, (i.e. cow-cow-cow, cowp, cowp, cowp) that tends to slow down in the end. They often announce their entry into an area with a loud barking call.

American Coot
Fulica americana

Other Common Names: Rafts
Spanish: Focha Cenicienta

American Coots are another chicken-like bird that are fond of marshes, edges of lakes, and canals or ditches. They are a plump dark-gray to black bird with a sloping white bill and forehead that has a small red patch on the forehead. They have little tails and short wings but large yellow legs and feet that make them strong swimmers. They can be up to 38 cm long. The young look like adults but are a bit lighter in color. American Coots are omnivorous diving birds that eat fish, tadpoles, insects, and vegetation. They are excellent swimmers but somewhat comical to watch maneuver on land. They require a long running take off over the water to get airborne. American Coots are another bird that a lucky scuba diver may see diving in Media Luna.

Photos by Steven Blandin and/Shutterstock.com

Belted Kingfisher
Megaceryle alcyon

Other Common Names: water kingfisher
Spanish: Martín Gigante Norteamericano

The Belted Kingfisher is one of the most magnificent birds to watch. They are usually seen perched over a body of waiting to dive into the water after their prey. One or two are almost always seen on the power lines over the irrigation canal on the road to Media Luna. Their favorite habitat is a body of water that is surrounded by trees and has vertically exposed earth for digging burrows where they nest. The clear waters of the Media Luna system are perfect for detecting their prey. Belted Kingfishers have a large slate-blue feathered head with a prominent crest. They have a thick and tapered bill, (that is longer than their head) that serves as an advantage when they dive head-first into the water after their food. They have a bright white collar around the neck and a dark band below. Males have a white underside. Belted Kingfishers are somewhat unique in that the female is more colorful than the male. Females have a reddish band below the dark band. Both sexes are about 32cm tall. Belted Kingfishers prefer eating fish and crayfish but will occasionally eat berries.

Photo by Bildagentur Zoonar GmbH/Shutterstock.com

Ringed Kingfisher
Megaceryle torquata

Other Common Names: Pied Kingfisher
Spanish: Martín Gigante Neotropical

The Ringed Kingfisher is larger than the Belted Kingfisher. Their body structure is the same as the Belted Kingfisher, but they are dark blue to bluish-gray with white markings and a white collar around the neck. They differ from the Belted Kingfisher by a rufous belly that covers the entire breast of the male. Just as with the Belted Kingfisher, the female Ringed Kingfishers are more colorful. Females have a bluish-gray breast and a white stripe between the breast and the belly. Ringed Kingfishers habitat is similar to the Belted Kingfisher's habitat. One big difference is the loud, sharp clattering voice of the Ringed Kingfisher. They are also divers and eat primarily fish, reptiles, and crustaceans.

Photo by Jo-anne Hounsom/Shutterstock.com

Northern Jaçana

Jacana spinosa

Other Common Names: Jesus bird, lily-trotters
Spanish: Jacana Centroamericana

When in flight, are medium-sized (17 – 23 cm long) wading birds that inhabit shallow marshes, canals, lake shores, and especially water edges with floating vegetation. They have dark brown bodies and a black head, breast, and neck. Their yellow bill has a white base followed by a yellow patch on the forehead. They have rufous plumage on their back, tail, and undersides. They are identified by their long legs and extremely long toes. One can see their yellow primary and secondary feathers when in flight. Males construct nests, incubate eggs, and do most of the work caring for the young. Juveniles have a white eyebrow and a few white feathers on the side of the head. Northern Jacanas eat insects, other invertebrates, and seeds that are picked up from the surface of plants or just below the water's surface. They also can use their feet to turn leaves over to get to insects on the undersides of leaves. Jacanas help to control insect pest populations.

Photos by Gail Johnson/Shutterstock.com

Osprey
Pandion haliaetus

Other Common Names: Sea Hawk, Fish Hawk, fish eagle, river hawk
Spanish: Águila Pescadora

Ospreys inhabit wetlands that contain fish and have nearby dead or open-topped trees. They are dark brown above and white below. Osprey's heads are white with a brown stripe from the eye to the back of the head. They are medium sized (50 – 66 cm) raptors. Females are usually larger than males. They have yellow eyes and a hooked black beak. The tail has alternating dark brown and white bands. Juveniles are duller in color and have a mottled appearance. Ospreys catch fish by hovering over waters to spot them and then diving into the water, feet first, where they reach out to their huge feet and use their talons to capture fish. One can plainly see an Osprey's fish catch held tightly in its talons as the bird carries it back to a nest or branch

Ospreys are observed around any body of water that is, at least, a meter deep and contain one of the larger species of fish in the Media Luna Watershed. They are usually sited in trees along the east canal and cypress trees near Los Peroles. Their stick nests are on top of electrical poles, cypress trees, and dead trees near bodies of water.

Mexican Duck
Anas platyrhynchos diaz

Spanish**:** Pato mexicano

Mexican Ducks, a subspecies of the Mallard, are dabbling (not diving) ducks and are found throughout the Media Luna Watershed. Both sexes of Mexican Ducks (*Anas diazi*) are similar and closely resemble female mallard ducks. Both genders are mottled brown with a bold eyeline and orange feet. Both sexes are 50 – 55 cm long and look a lot like female mallards but with a somewhat darker body. They have yellowish/green or olive bills, although the males are usually brighter than the females and yellow. The female's bill sometimes appears green/orange. Mexican ducks are mottled brown with a blue patch edged with white on the wing, noticeable in flight or at rest. They have a dark eye line and orange feet. Juveniles are a dull, mottled brown duck. Mexican Ducks are foragers and will eat a wide variety of food. They dabble (tipping forward in the water to eat seeds and aquatic vegetation) to feed rather than diving. They eat aquatic insect larvae, worms, freshwater shrimp, and snails. Mexican Ducks are easy to domesticate and can be seen wandering around Media Luna's sales booths where they are happy to accept a handout.

Photo by G Tally/Shutterstock.com

Photo by Jim Conrad

Green Kingfisher
Chloroceryle americana

Spanish: Martín Pescador Verde

The Green Kingfisher is smaller than the other kingfishers and is absent of the blue-gray coloration. They are small compared to the other kingfishers, averaging 17 to 22 cm in length with a long dark bill, up to 50 mm long. Most males are dark green over their dorsal parts and head with a broad white collar. The ventral parts are white with a chestnut patch on the breast. Their wings are slightly mottled with white spots. The tail is dark green with white outer feathers seen while in flight. Females are smaller and lack the chestnut patch but have two spotted green bands on the breast. Bothe males and females have a small crest and black legs.

Green Kingfishers like more inaccessible areas where there is plenty of overhanging brushes to provide perches for hunting. At Media Luna, they may be located along the north canal that is only accessible by canoe or kayak. They are often seen along the irrigation canals where the brush grows up to the shoreline, especially the canals flowing from Los Peroles. Green Kingfishers hunt for prey from a low waterside perch, looking for crustaceans, freshwater prawns, aquatic insects and especially small fish.

Muscovy Duck

Cairina moschata
Other Common Name: backyard duck, mute duck
Spanish: Pato Criollo, Bragado, Pato negro, Pato Real, Pato real o negro

The species, *Cairina moschata*, is separated into two subspecies, the wild subspecies, *Cairina moschata sylvestris*, and the domestic subspecies, *Cairina moschata domestica*. Our concern will be with the local subspecies because they are abundant at Media Luna. Although the wild species, *Cairina moschata sylvestris*, is native to Mexico, they do not inhabit the Media Luna Watershed. The domestic subspecies is heavier than the native subspecies and are not able to fly long distances.

One of the most distinguishing physical traits of all Muscovy ducks is a bright red, bumpy, and featherless mask around the eyes and beak. The mask is larger in the male. Most Muscovy ducks have a "crest" but the males are much larger to make an impression on females. The wild subspecies are sleek black/brown with small patches of white. The domestic subspecies can be any variation of black, white, brown and even blue or calico. Some adults at Media Luna are nearly entirely white. Adults weigh up to 7kg and a length of 75 cm. Males are larger than females. Because of their large size, Muscovites are sometimes thought to be geese. Their feet are large with long sharp claws for holding onto tree branches. Even though they have webbed feet, they are not commonly seen swimming because they have under-developed oil glands.

Muscovies found around Media Luna are semi-tamed. How they first arrived at Media Luna is questionable. Because of their eating habits, they are thought as being both pests and helpful birds. It is a fact they consume foods dropped on the ground that would usually attract rodents. Muscovies also eat insects such as roaches, flies, spiders, and mosquitoes. They will also consume mosquito larva in the water, functioning as a means of pest control. On the other hand, Muscovies are constant and belligerent beggars for handouts while leaving their droppings in inappropriate places, such as the lake, canals, and channels. Droppings could result in a possible health problem or, at least, questionable water quality. They also eat young new plants, something that is becoming rare around Media Luna. Domestic Muscovies are also known to transmit diseases to wildfowl. Another problem observed at Media Luna is their ability to interbreed with native wild ducks, "muddying up" the gene pools.

Chapter 8

REPTILES AND AMPHIBIANS OF THE MEDIA LUNA WATERSHED

We do not inherit the earth from our ancestors; we borrow it from our children.
Native American Proverb

Our lists of reptiles and amphibians are based on our personal findings and on the following reference where specimen records were provided:

"Lemos-Espinal, Julio A. and James R. Dixon. 2013. Amphibians and Reptiles of San Luis Potosi, Eagle Mountain Publishing, Eagle Mountain, Utah. 300 pp."

REPTILES

Turtles, Snakes

Mesoamerican Slider
(*Trachemys venusta*)
Spanish: Pecho de carey - Tortuga de Guadalupe

Mesoamerican Sliders are the most widespread turtle in North America. It is also the most common pet turtle in North America. Mesoamerican Sliders are native to Midwestern and Southern areas of the United States and Northern Mexico but has become established in many other places around the world because of pet releases. It is considered an invasive species in many areas where it is not native. The Mesoamerican Slider is native to the Media Luna Watershed.

Mesoamerican Sliders have a unique distinctive red or orange (sometimes yellow) stripe behind the eyes. They get their name, "Slider," from the way they seem just to slide off rocks or logs when disturbed. They can grow up to 25 cm long and 25 cm wide. The yellow-green to dark green carapace can have patches of off-white, yellow, or even red and is relatively oval and somewhat flat. The plastron is yellow with light and dark patchy markings. The legs, tail, and head are green with thin, yellow random lines. Although males are a bit smaller than females, they have thicker and longer tails. Identification troubles can occasionally happen because the Mesoamerican Slider sometimes interbreeds with other sliders.

After mating, a female may lay between 2 and 20 eggs in a nest dug in the earth on land in a sunny area. It is critical to the Mesoamerican turtle that the shorelines along the undeveloped areas remain available for egg laying. Eggs will hatch in 60 to 90 days after being laid. Hatchlings break their egg shell with what is called an egg tooth. The egg tooth breaks off after a couple of hours following hatching and never grows back. Many different animals eat the hatchlings to include: domestic and wild cats, dogs, large birds, large fish, fox, snakes, squirrels, and raccoons.

Mesoamerican Sliders are a semi-aquatic turtle that prefers calm, fresh, and warm water. Therefore, Media Luna is a perfect habitat for them. Red-Eared Sliders need an area where they can bask in the sun (i.e. logs, rocks, or flat land). They also require an abundant amount of aquatic vegetation because it is the central part of an adult's diet.

The observed decrease of the Mesoamerican Sliders at Media Luna is most likely due to the loss of habitat and being captured for pets. Although no scientific studies have been done to demonstrate a loss of population at Media Luna; scuba divers have seen fewer and fewer turtles in the water over the past few years.

Photos by Matt Jeppson and Brian Lasenby/Shutterstock.com

Spiny Softshell TurtlePhotos

(Apalone spinifera)

Other Common Names: Texas Spiny Softshell, Gooseneck Turtle,
Leatherback Turtle
Spanish: Tortuga-casco suave espinosa

At one time, Media Luna may have been the southern-most region for Spiny Softshell Turtles. Scientific references that included San Luis Potosi showed the Spiny Softshell Turtles range did not contain San Luis Potosi. We know they are there, (or were there) because scuba divers have observed many over the years. Although, since Media Luna has become a State Park, the Spiny Softshell Turtles have disappeared from the lake.

Spiny Softshell Turtles are olive green to brown with black spots and a soft leathery carapace with small projections on the edges. The carapace also lacks scales and scutes and is somewhat flat. The males shell has a slight sandpaper feel to the touch. The adult female shell is flat, but there are several significant spine-like projections at the border of the carapace.

The plastron is also soft. It is lighter in color compared to the carapace. The neck and head are long and flexible with a tapered snout (hog-like). The large webbed feet are brown/green forming broad paddles. They can grow up to 38 cm long with females being larger than males and with lighter and smaller tails. Males are likely to maintain their bright coloration while females tend to get darker with age.

Spiny Softshell Turtles are found mostly in slow-moving rivers, streams, ponds, and lakes. They are diurnal and entirely aquatic although they do enjoy basking in the sun on logs and rocks. Although, the female will dig nests on land during the spring and summer and lay one or two clutches of up to 33 eggs. They enjoy lingering unseen underwater with the snout poking up to the surface to breath. They are primarily carnivorous, eating crayfish, worms, snails, fish, frogs, tadpoles, and other reptiles. They are also known to scavenge for food.

Spiny Softshell Turtles can live up to 50 years. They are difficult to get near to as they scare easily and can move very fast on both land and water. Many people get seriously scratched or bitten when handling a Spiny Softshell Turtle. They are often caught at maturity and eaten by man. They also are regularly caught and kept as pets.

Mud Turtles
Mexican Mud Turtle (Guanajuato mud turtle)
Kinosternon integrum

Spanish: tortuga-pecho quebrado mexicana

According to van Dijk, P.P. et al., there are two species of the Mud Turtles in the Rio Verde, the Rough Footed Mud Turtle (*Kinosternon hirtipes*) and the Mexican Mud Turtle, (*K. integrum*). While the Mexican Mud Turtle is native to the watershed, the Rough Footed Mud Turtle is a subpopulation introduced into the State of San Luis; probably through the pet trade. At the time of writing this book, the author has only seen the Mexican Mud Turtle.

Mexican Mud Turtles prefer slow-moving fresh water with deep pools. In the Media Luna Watershed, they are usually seen near the shorelines. Mexican Mud Turtles are more active during summer months. Mexican Mud Turtles are often considered aquatic in nature but spend a fair amount of time out of the water. They prefer pools with large amounts of aquatic plant material. Mexican Mud Turtles are small turtles that grow to a length of 20 cm. The external shell or carapace is somewhat round, smooth, and in various tones of brown. The plastron (underside) is yellow and usually has 11 scutes (sections). The head is usually dark brown on top and lighter on the sides and bottom. The body's skin is relatively smooth. Male turtles are larger than females and have a larger tail. They have long claws and webbed feet. A noteworthy characteristic of all mud turtles is their hinged shell. The underside has a hinge at both the front and back and a fixed one in the middle. The hinges aid the turtle with being able to pull its soft parts into its body for protection. With the hinges, it can close its shell completely.

Mexican Mud Turtles hunt for food by strolling on the bottom of a shallow pool or channel where they consume invertebrates and plants. Being both omnivorous and carnivorous as well as opportunistic, mud turtles will eat just about anything they find. They may lay several clutches of one to five eggs per year. Incubation time is three to five months. The local weather conditions play a big part in determining the incubation times. Mexican Mud Turtle populations have decreased during the past few years due to visitors collecting them for pets.

Photo by Seth Patterson

SNAKES

Snakes are a common reptile in the Media Luna Watershed. Most are land species and often seen along the various canal or channel shores. These species include Jan's Night Snake, (*Hypsiglena jani*), Milk Snake (*Lampropeltis triangulum*), Great Plains Rat Snake (*Pantherophis emoryi*) Tampico Thread Snake (*Rena myopica*), and the genus of venomous pit vipers called Crotalus (rattlesnakes). We will discuss the one species that truly depends on the water for food and shelter, the Diamond-backed Water Snake (*Nerodia rhombifer*).

Photos by Charlene Key/Shutterstock.com

Diamondback Water Snake
Nerodia rhombifer

Spanish: Culebra De Agua De Diamantes

80

The name Diamondback Water Snake (semi-aquatic) is a bit misleading because they do not have distinctive diamond markings down their body. Instead, the markings have a reticulated (to mark with lines resembling a network) pattern. Vaguely diamond shaped blotches are in the middle of the back with bars that stretch up their sides. Adults are brown or olive green with the brown or grayish spots. Their undersides are somewhat yellow or a lighter brown and sometimes with black blotching. They grow to a length of 1.5 meters but can reach 2.5 meters in favorable environments. Young diamondback water snakes appear like light versions of the adults.

Diamondback Water Snake females produce eggs that hatch within their body. They breed in the spring and give birth to 20 or more 10-25 cm young in late summer. They eat mainly frogs and toads but slow moving or stranded small fish and crayfish. When possible, carrion is also consumed. When handled, they will viciously bite and repeatedly defecate on the handler. Diamondback Water Snakes are non-venomous but very aggressive and often misidentified as poisonous cottonmouths. In the Media Luna Watershed, Diamondback Water Snakes are often seen basking on logs and brush near water.

Amphibians

Gulf Coast Toad
Bufo valliceps / Incilius nebulifer

Other common names: Coastal Plains Toad, Common Toad
Spanish: Sapo Común

The Gulf Coast Toad is a flat black/brown toad with a white or yellowish stripe that goes down its back and sometimes lighter patches along the sides. Their undersides are cream colored. They range from 5 to 13 cm in length. They have wart-like tubercles on their back but none on the underneath. Gulf Coast Toads have well-developed cranial crests that form a valley between them and go from the nose to the back of the head and wraps around the back side to the eyes. They have a well-developed triangular shaped parotoid gland behind the eye that looks like a wart. One can tell the difference between a male and female by checking out the color of the throat: males are yellow-green, and females are white.

The Gulf Coast Toad is common in many different habitats within the Media Luna Watershed. They have most often been seen crossing roads following rainfall. They prefer areas where there is permanent water because they need the water for breeding. They are sometimes found gathered under street lights to feed on insects that are attracted to the light. They are known to travel long distances while searching for food. The Gulf Coast toad is a carnivore that eats small insects. They are most active a twilight.

Around the Media Luna Watershed, males can be heard singing from March to September. They produce sounds that last about 4 seconds and mimics a wooden rattle. They are often heard following rainfall. The call corresponds with the breeding season. Females lay their eggs in extended threads and often enclosed around the foliage. Their eggs are found along the shores of the channels where there is no human activity as well in surrounding temporary ponds or other bodies of water, especially following rainfall.

Photo by Karon Troup/Istock.com

American Bullfrog
Lithobates catesbeianus (Rana catesbeiana)

Other common names: Bullfrog, Croaker
Spanish: Rana Toro

The American Bullfrog is believed to be an introduced species to the Media Luna Watershed. They are known for their baritone "jug-a-rum" call that some people think sounds like a cow mooing. The males are the only ones that make this call, heard during the daytime and evening. The American Bullfrog can grow to a length of 20 cm (8 inches) and weigh up to 750 grams (1.5 lbs.), and males are usually slightly smaller than females. They are the largest frog found in North America. Their color ranges from olive-green to a gray-brown and has easily distinguishable round eardrums on the sides of their head. They often have dark brown spots. Their bellies are white to yellow and sometimes with black spots. During the breeding season, the throat of the male is yellow, and the females are white. The forelegs are short and robust, and the hind legs are long. Their back feet are fully webbed.

American Bullfrogs live near water. The only area the author has heard a bullfrog within the watershed is along the east channel where they can find warm, calm, and shallow waters. They can live in both shallow or deep water. They are nocturnal predators that will eat anything they can fit in their mouths including birds, fish, snakes, small turtles, other frogs, small mammals, crayfish, and insects. Males are known to be extremely territorial and will aggressively guard their territory especially during the breeding season. Their breeding season generally lasts two to three months. The season of reproduction begins May and continues through July. Within the basin, the making season often lasts longer. Females lay thousands of eggs during the breeding season. Complete metamorphosis usually lasts around 80 days in the area of the watershed

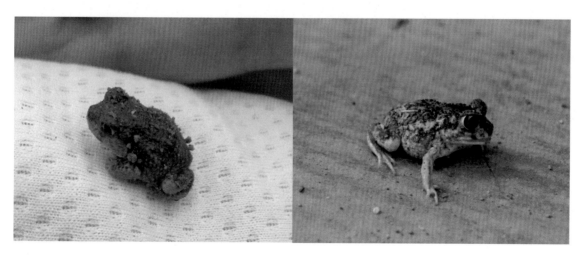

Mexican Spadefoot
Spea multiplicata

Other common names: Mexican Spadefoot Toad, Desert Spadefoot Toad, Southern Spadefoot Toad
Spanish: sapo spadefoot Mexicano

The Mexican Spadefoot, *Spea multiplicata,* is brown, grayish-green, or grayish-brown, with black blotches on its dorsal parts. They sometimes have light stripes on their backs or sides. One of the most prominent characteristics of the Mexican Spadefoot is their large eyes that give this toad a somewhat comical appearance. They also have horny wedge-shaped tubercles used for digging on the bottom side of each hind foot. Large adults are around 4 to 7 cm long. They do not have a boss (raised bony area on a toads head between the eyes) like other spadefoot species.

The females will lay nearly 1000 eggs attached to vegetation. The eggs incubate for only one or two days before hatching. Depending on environmental conditions, the tadpoles will complete metamorphosis in 15 to 45 days. Tadpoles have a broad, flat head and can grow up to 6 cm in length although most tadpoles complete metamorphosis before reaching 3 cm. The tadpoles come in two morphs, 1) carnivores, and 2) omnivores.

The Mexican Spadefoot can remain buried in the soil for months and then emerge only during rainfalls. They emerge from the ground to breed as soon as the summer rains begin. Their preferred breeding sites within the watershed are along the canals and channels where water collects and forms small pools. They are nocturnal and will remain active as long as the soil remains moist. The Males call while floating in the water. The call sounds like someone running their fingernail along the teeth of a comb. When handled by humans, adults emit a secretion that smells like peanuts.

Right photo by Matt Jeppson/Shutterstock.com

Rio Grande Leopard Frog
Lithobates berlandieri

Also called the Mexican leopard frog

The Rio Grande Leopard Frog lives along the channels and canals as well as the lesser used areas around the lake. This species can grow up to 5.7 to 11.4 cm long. They use both permanent and temporary water. They have dark spots on a background of green or brown or a combination of green and brown. Their ventral surface is cream-colored. They have long powerful legs and webbed feet. Their noses are angular and often absent of spots. Tadpoles may reach up to 9.5 cm in length and are light-colored and mottled.

The Rio Grande Leopard Frog is mostly nocturnal although they are often active during the day. They prefer habitats with standing water in arid or semiarid areas. They are insectivores but are known to eat almost anything they can swallow. They mate during rainy periods in the spring and early summer and lay their eggs in sizeable masses fastened to aquatic flora. Eggs and larva develop primarily in non-flowing water.

Throughout the night, males make an advertisement call during mating season. Their call is a short, low-pitched trill or rattle, usually lasting less than a second and given singly or in rapid sequences of two or three trills. The advertisement call of the Rio Grande Leopard Frog can be heard up to a quarter of a mile away. Males make the calls around bodies of water to attract females, as well as a warning to other males of their presence.

Chapter 9

CRUSTACEANS OF THE MEDIA LUNA WATERSHED

Crustaceans belong to the same phylum as insects, "Arthropods." They have segmented bodies, jointed limbs and a hard exoskeleton that protects the animal from predators and aids in preventing water loss. As the animal occupying the exoskeleton grows, the crustacean molts and forms a new exoskeleton. The new exoskeleton is somewhat soft but hardens over time. It is during the soft stage that the animal is most vulnerable to predators. Most have a distinct head, thorax, abdomen, and gills for breathing.

Crustaceans in the Media Luna Watershed include three crayfish and a freshwater shrimp. Most records of crustacean captures in the watershed do not explicitly state where the specimens lived. Although, it was recorded in 1972 that 500 crustaceans came from the canals and springs of Ejido Las Palomas. The photos were all taken by Seth Patterson. Perhaps this is the first time photos of this quality of these three species have ever been published from the Media Luna region.

Photo by Seth Patterson

La Media Luna Crayfish
Procambarus (Pennides) roberti

The La Media Luna Crayfish in a native to the basin. We were able to locate this crayfish at six different locations in the lakes and throughout the canals and channels. One observation was that there was a substantial increase in harvesting crayfish for human consumption. It is evident that monitoring is required to determine whether harvesting and habitat degradation is resulting in significant population declines. Some organizations consider this crayfish as threatened while others think it endangered. The populations are substantially decreased during the times the canals are cleared of vegetation.

They are reddish/purple dorsally and red/orange ventrally. Their claws are black dorsally and with red/orange tips. The entire carpus appears red/purple when first removed from the water. These crayfish are often seen in the canals but harvesting them may jeopardize their survival.

Photo by Seth Patterson

Australian Red Claw Crayfish
Cherax quadricarinatus

Other Common Names: Queensland Red Claw, Redclaw, Tropical Blue Crayfish, Freshwater Blueclaw Crayfish

The Australian Red Claw Crayfish, *Cherax quadricarinatus* was first introduced into Mexico during the 1990's for aquaculture purposes. Australian Red Claw Crayfish first occurred in the canals near Media Luna in 2012. Today, it is harvested from the canals for human consumption. They are much larger than native crayfish and have an excellent flavor. No one knows exactly how they entered the watershed, but it is assumed that the first ones were deposited into the waters by aquarium hobbyists.

Australian Red Clay Crayfish are beautiful animals. They are relatively large freshwater crayfish with red with maroon highlights on a blue-green body. Mature males have bright red streaks along the margins of their large claws. Males can weigh up to 500 g and females up to 400 g. They are scavenging omnivorous and opportunists. Juveniles consume zooplankton. Typical water rats consume them along with water birds, especially cormorants.

There is concern that native endemic crustacean species may be displaced by this invasive species through modification of natural habitats, direct competition, predation, and its potential to transmit new parasites. Like other freshwater pests, (i.e. tilapia), once populations endure, extermination is virtually impossible. Regrettably, there are insufficient studies on the ecology of Australian Red Claw Crayfish.

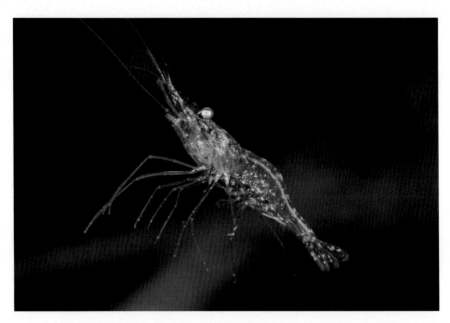

Photo by Seth Patterson

Freshwater Shrimp

Freshwater Shrimp - *Palaemonetes lindsayi,* are a critically endangered species found in the Media Luna Basin. This species is only known to exist in the Media Luna Basin. They are known to be threatened by sunscreen polluting the water, heavy water extraction from the aquifer for urban consumption, and the introduction of introduced species such as the Austrailian crayfish and Tilapia.

The average visitor would not even know this species exists because they are so difficult to see. They are mostly transparent. One would have to snorkel around the shorelines to get a good view of them. They live in the shallow waters close to shore.

Chapter 10

ENVIRONMENTAL IMPACTS

"Education is the most powerful weapon that you can use to change the world."
Nelson Mandela

Ecology is the branch of science concerned with the interrelationship of organisms and their environments. Ecology includes both living and non-living parts of an ecosystem. The ecosystem is made up of all the living plants, animals, bacteria, rocks, water, and even the atmosphere. It is important that we understand this simple fact to manage any changes we make to the ecosystem. The ecology of the Media Luna Watershed includes the relationships between humans and all the living creatures that exist in the entire watershed as well as humanity and the non-living, such as minerals in the soils and water. Ecology deals with the balance of nature, the balance between man and his environment. Barry Commoner, an American Biologist, University Professor, Ecologist, and Politician stated it best when he said this about ecology, **"The first law of ecology is that everything is related to everything else."** This natural law is an important concept as we discuss environmental impacts. Also, keep in mind that each native species of plant and animal plays a critical role in the grand scheme of maintaining a healthy ecosystem.

So what happens when one of the ecological factors changes? The answer is; we have an environmental impact. Environmental impacts may be beneficial or adverse. Environmental impacts are often associated with economic development projects like the development of Media Luna into a major recreational park. Environmental impact, by definition, involves a change in the ecosystem conditions and a new set of unfavorable or beneficial environmental consequences.

Environmental impact identification is essential for proper management of an ecosystem. There are several possible impact categories, such as:

- Economic/ Social
- Cultural
- Physio-Chemical
- Health
- Biological

ECONOMIC/SOCIAL IMPACTS: The development of the park and the continuous use of the Media Luna Watershed has been a tremendous economic boost to the surrounding communities. However, let's first discuss the most significant impact on the economy and social structure, "irrigation." Over 15,000 hectares of crops are irrigated annually. The crops include corn, citrus fruits, peppers, tomatoes, beans, and sweet potatoes. The irrigation canals and channels are also a valuable source of water for livestock. Irrigation has resulted in an improved standard of living for those who can take advantage of it. The number one industry in the entire Rio Verde Valley is agriculture, thanks to the Media Luna Watershed.

The creation of "Parque Estatal de la Media Luna" has contributed to an improved economy for the local communities. The park is responsible for hundreds of jobs throughout the local commerce. Jobs are created and secured for more than just the employees that work at the park; whose numbers include booth sales, maintenance, dive masters, security, and managers. We must also consider the labor required for delivering food items to Media Luna, bus drivers who provide visitors, the farmers who grow the crops sold at the park. Also affected is the local commerce that keeps the park going such as hotels, restaurants, dive shops, and retail sales.

The Ejido receives money from the entrance fees of visitors and the booth operators. They may use this income for community development and improvements. The economic impacts strongly influence the social consequences.

CULTURAL IMPACTS: Media Luna is a hotbed for cultural artifacts. The artifacts have lead to an enhanced understanding of local history. For many years, visitors have been allowed to remove historical artifacts from Media Luna. Luckily, some of the artifacts have been officially registered and preserved in several museums. Today, looting of underwater artifacts is monitored through the direction of the Ejido.

PHYSIO-CHEMICAL IMPACTS: This category includes both physical and chemical properties. Impacts disturbing soil and landforms, floods, and sedimentation are considered physical effects. Activities that result in a chemical change of land, water, or air are considered chemical impacts.

Luckily, the activities at Media Luna contribute little to the physio-chemical conditions. There isn't any known intentional industry waste deposited into the lake, channels, or canals. Agricultural practices can cause an adverse impact on water quality with the improper use of fertilizers and pesticides. There is certainly runoff and seepage of agricultural residue into the Media Luna Watershed.

HEALTH IMPACTS: These effects are determined by a range of factors resulting in changes in the economic/social status, cultural adjustments, physio-chemical and biological conditions within a population. The question in this situation is, "How are these changes affecting the health of the community?" Health impact assessments are multidisciplinary processes whose aim is to assess the health consequences on a particular population, in this case, the people living in the area of the Media Luna watershed.

BIOLOGICAL IMPACTS: Biological impacts include living matter such as wildlife, vegetation, and aquatic life. The biological category includes the functions and interactions of all components in an ecosystem. Thus, we return to the ecology affected by a change. Now is when the management of an ecosystem can become complicated because someone has to determine where the balancing line between all the categories of impacts. The biological scheme interacts with all the other impact categories, each of which can impair the balance of nature. Human activities, such as camping, building fires or making trails are all done to benefit one's personal needs. Unfortunately, they often damage the natural balance and lead to the manmade destruction of a vital link in the ecosystem.

At Media Luna Park, the environmental management organization responsible for determining the balance between environmental impacts is the State of San Luis Potosi while the Ejido has the responsibility for daily management. The coordination between these two entities is essential for proper environmental management to reduce biological impacts. The State has issued many important rules at the park for visitors to follow: rules designed to control some of the environmental impacts. Enforcement of the rules is improving over time but needs to be vigorously enforced.

EXAMPLES OF ENVIRONMENTAL IMPACTS

Let's review some widespread impact cases observed at Media Luna. While reading about each impact, consider the categories each effect encompasses.

EXAMPLE 1: Overexploitation

Overcrowding is a result of overuse and is a major problem at Media Luna. On holidays, as many as 6,000 people may visit the park. Overcrowding results in visitors using environmentally sensitive areas for camping, hiking, and picnicking. This activity results in the destruction of native vegetation, primarily due to compaction of the soils, otherwise known as trampling. Too many tourists using the same areas over and again trample on the undergrowth and topsoil where they ultimately damage and change the ecosystem resulting in:

1. Loss of biodiversity
2. Fracture and bruising of plant stems
3. Reduced plant development
4. Shortage of organic matter in soils
5. Decrease of top soil microporosity
6. Surge in runoff from tourist areas into natural aquatic habitats
7. Increased of erosion

Overcrowding also encourages visitors to travel off the established paths and to wonder into natural areas not intended for tourist camping or hiking. When prescribed trails and camping sites fail to give guests the access and experiences they desire, visitors regularly venture "off the trail" to reach locations not accessible by prescribed trails. Reasons can be to view new surroundings, avoid crowds, or simply to explore. There is a greater need to wonder off the path when there is overcrowding.

However, for the local community, overexploitation brings more money into the community resulting in a positive economic/social impact. So now, someone needs to make the balancing decision between economic and biological impacts? The bottom-line comes down to "who has control at the gate" and to where visitors are allowed into the park. The Ejido has control at the gate so naturally they let as many visitors into the park as they desire, after all, it means more income to the Ejido. So the balance favors the overexploitation of the park. Critical limits need to be established and enforced.

EXAMPLE 2: Uncontrolled Tent Camping

In the past, if visitors wanted to tent camp, they could place their tent almost anywhere desired "until recently" even though there are signs with rules identifying where tents are allowed. We often see campers putting up their tents in the few areas there is a groundcover, Again, we have a situation where there needs to be a decision between economic and biological impacts. Enforcing the rules for where a tent placement is apparently taking away the most convenient sites. The most suitable sites are those close to the swimming area and in the shade of the giant cypress and Australian pine trees. They are also closer to the sales booths and restrooms.

Media Luna has made enormous strides towards improving this situation, "on paper." Enforcement of the rules is dubious. Today, tent campers are required to place their tents outside the area of the Cypress and Australian pine trees to allow the understory to recover to its natural condition. Now, campers are putting their tents in the few areas where there is still vegetation. The environmental changes taking place from campers setting tents in sensitive areas are; obliteration of new vegetative growth, soil compaction, and the establishment of new fire pits. Currently, many campers are allowed to place their tent around environmentally sensitive vegetative edge communities rather than in the area designated for tents.

Example 3: Biodiversity Decline

Media Luna is a rare oasis of life due in part to its biodiversity. With the loss of biodiversity, the web of life that surrounds Media Luna is threatened of being reduced, species by species. Over the last ten years, there has been a need for more space for camping, picnicking, swimming, and playing; this has resulted in damage or alteration to the natural habitats that support the biodiversity. No one knows how many species are required to maintain a healthy environment, but recent scientific findings indicate that the healthiest environments are those with the greatest biodiversity. Therefore, the greater the biodiversity, the greater the chance of species' survival. Maintaining the natural biodiversity is the first step in preserving Media Luna.

Let's take, for example, what appears to be a small change at Media Luna. Keep in mind that looks can be deceiving. This case may seem to be on a small scale and not look like much, but if one adds up all the little effects over time, the result could be immense. One change that has occurred over the last few years is the loss of what once was a healthy population of leaf-cutting ants. Most people would say, "So what, they are just ants!" Well, leaf-cutting ants once inhabited the park's understory and played a significant role in maintaining the ecological balance of the park. Leaf-cutting ants aided in the maintenance of productive soil conditions around and within the campgrounds of Media Luna. Today, the ants are gone. Where trails of leaf-cutting ants used to be, are tents and trails surrounded by densely compacted bare ground.

A leaf-cutting ant's nest can be as large as the size of a car. Nest refuse and soils associated with leaf-cutting ants have a greater concentration of macronutrients and soil penetrability about non-nest soils. Therefore, the soils around Media Luna used to be more aerated with the excavation of tunnels and chambers and enriched with more nutrients as the ants bury their waste products and produce a nutrient-rich fungus. This natural recycling of nutrients in the environment no longer takes place.

Example 4: Uncontrolled Fire Pits

Figure 20 - Fire pits like this one cause severe damage to the natural environment.

We all love open fires when camping. Open fire pits on the ground are excellent at the moment. However, consider the long-term effects that include damaging trees for firewood, burning tree and plant roots, wood smoke pollution, unsightly scars on the ground, and sterilized soil preventing vegetation from germinating for many years.

Example 5: Loss of Habitat

Figure 21 - The irrigation canal that flows north of the springs becomes abundant with aquatic vegetation that supports thousands of crayfish, turtles, and fish.

Every few years, the Rio Verde Irrigation Authority cleans out the 11.5 km north canal, thus killing the fish, crayfish, aquatic plants, insects, and turtles that inhabit the canal. The practice applied is similar to clear-cutting a forest; if it is alive, kill it. One supposes this is done to improve the water flow. In this case, looks can be deceiving. The canal looks beautiful and pristine with no vegetation in it, what-so-ever, giving the viewer a sense of purity. However, consider the extermination of endemic fish, including endangered species such as the Checkered Pupfish. Required repairs for leaks will take place, but the need for destruction of the entire canal ecosystem to fix a leak is questionable and more important "preventable." Possibly, cleaning the canal of vegetation in sections on alternative years may preserve some of this precious ecosystem and its inhabitants.

Example 6: Introduction of Invasive Species (Invasive exotics)

The introduction of exotic invasive Species may be the greatest environmental impact on the Media Luna Watershed, especially on Media Luna itself. The non-native Australian pine (*Casuarina species*) dominates the canopy surrounding Media Luna while tilapia (*Oreochromis sp*) school in the lake, commanding prime nesting areas that native fish require. Other species of exotics include non-native evergreen tree species, Australian red claw crayfish, (*Cherax quadricarinatus*), Rio Grande Cichlid, (*C. cyanoguttatum*), Panuco Gambusia, (*Gambusia panuco*), Broadspotted Molly, (*Poecilia latipunctata*), Shortfin Molly, (*P. Mexicana*), red-rimmed melania, (*Melanoides tuberculate*), giant reed, (*Arundo donax*), North American bullfrogs, (*Lithobates catesbeianus*), bermuda grass, (*Cynodon dactylon*), common pigeons (*Columba livia*), house sparrows, (*Passer domesticus*), eucalyptus, (*Eucalyptus globulus*), and castor oil plant, (*Ricinus communis*). I'm sure many more exotic and invasive species could be found on a survey specifically designed to locate exotics. Potential invaders are American crocodiles, nutria, aquarium fish, water hyacinth, carp, hydrilla, Northern snakeheads, zebra mussels, various species of catfish and Elephant Ear plants. As the watershed continues to grow, more exotics will be introduced threatening the native species (many endemics) to the point of their extinction.

Figure 22 – Aquatic plants covered with silt.

Example 7: Degradation of Aquatic Vegetation

One of the most obvious environmental impacts has been the decline of aquatic vegetation, especially the water lilies, *Nymphaea* sp. Swimmers have destroyed most of the plants by kicking and trampling on them. Another serious problem is when swimmers step into the detritus on the lakes bottom causing particles to float to the surface and then eventually sinking and landing onto the leaves. The detritus particles shade the leaves resulting in less photosynthesis. The leaves eventually turn yellow or brown and sometimes dies.

Where to Go from Here?

One limitation of environmental management at Media Luna is the lack of scientific information available for evaluating biodiversity loss rates and habitat tolerance to human impacts. Each ecosystem has its characteristics, resources, and species' distribution. Lack of reliable baseline data and a scarcity of qualified field workers make it harder to assess the situation. An excellent starting point would be a robust population and diversity baseline. A population and diversity baseline would be a valuable tool for evaluating environmental changes, (environmental impacts).

Another dilemma occurs once an aquatic ecosystem becomes damaged; it is tough to restore it to its natural condition. For example, consider the overpopulation of tilapia in Media Luna. This exotic fish species has become an invasive species and is responsible for several noticeable changes in the lake (i.e. species' distribution and water turbidity). How does one go about eliminating only one species in a lake consisting of eleven other species, especially when the target species has become the dominant species while some of the other species are threatened or endangered? The continued population growth of tilapia in Media Luna could change the natural food web resulting in the decline of biodiversity.

It is imperative that all governmental agencies involved with managing the watershed, as well as local citizens, assemble as a team for the purpose of gathering essential ecological information for decision makers to analyze. Only then will environmental changes and problems, either potential or existing, be identified and sensible resolutions achieved. Educating the local citizens is essential. Along with education about the watershed will come pride and a need to preserve the natural treasure they own.

It may be wise for the planners of Media Luna to review the steps that have been taken at, Aquarena Springs, Texas (now known as the Meadows Center for Water and the Environment) http://www.austinchronicle.com/news/2008-11-14/702485/, where planners have made many changes towards saving the natural environment of a water system similar to the Media Luna Watershed. They have restricted swimming in most areas and established a unique course for scuba divers; all in an effort to eliminate the negative environmental effects man has on the lakes ecosystem.

Glossary

Achene – A small, dry, one-seeded fruit that does not open to release the seed.

Algae – A group of aquatic single or multi-cellular organisms that contain chlorophyll but lacks roots, stems or leaves.

Apex Predator – A predator that is at the top of a food chain.

Aquifer – a layer of rock or sand that can absorb and hold water.

Artesian spring – A spring whose water is under pressure and reaches the surface through some fissure or another opening in the confining bed that overlies the aquifer.

Autotrophs – An organism capable of making its food from simple inorganic substances using light or chemical energy.

Bioindicator – Organisms whose standing in an ecosystem is scrutinized as a sign of the ecosystem's heath.

Biological oxygen demand – The quantity of oxygen needed by aerobic microorganisms to decompose the organic matter in a sample of water.

Boss – A raised boney area on a toad's head between the eyes.

Canal – For the purpose of this book: a manmade waterway constructed to convey water for irrigation.

Carapace – A hard or chitinous case or shield covering the back or part of the back of an animal such as a turtle or crab.

Channel – The area where a natural stream of water runs.

Decomposers – Organisms such as a bacterium, fungus, or invertebrate, that have the ability to decomposes organic material.

Detritivores – An organism that feeds on dead organic material, especially plant detritus in the Media Luna Lake.

Dissolved oxygen – Minute bubbles of gaseous mixed in water and available to aquatic organisms for respiration. Sources of DO comprise the atmosphere and aquatic plants.

Electrical conductivity – The measure of the amount of electrical current a material can carry. For the purpose of this book, it refers to the amount of electrical current can carry in water.

Emergent Plants – A plant that grows in water but pierces the surface so that it is partly in air.

Environmental Impacts – For this book, possible adverse effects to the environment caused by human activity.

Fecal bacteria – Bacteria whose usual habitat is the colon of warm-blooded animals.

Floating plants – Plants that grow in water whether rooted in the mud or floating without anchorage.

Food web – A scheme of linking interdependent food chains.

Headwaters – The source of a stream or river.

Hydrologic cycle – Also called the water cycle. The natural sequence through which water passes into the atmosphere as water vapor precipitates to earth in liquid or solid form and ultimately returns to the atmosphere through evaporation.

Invasive species (exotic) – A species that comes from a distant region and has the potential to cause significant damage to a local ecosystem.

Invertebrates – Animals without a backbone.

IUCN – International Union for Conservation of Nature. The IUCN Red List is the world's most comprehensive inventory of the global conservation status of plant and animal species. It uses a set of criteria to evaluate the extinction risk of thousands of species and subspecies.

Karst Springs – Kast Springs are characterized by a network of conduits and caves formed by chemical dissolution. The widespread dissolution of rock leads to the development of subterranean channels through which groundwater flows.

Lentic – Still waters such as lakes, ponds, or swamps.

Limnetic zone – The well-lit, open surface waters in a lake or pond away from the shore. An area that light penetrates in the open water.

Littoral zone – The shallow vegetated zone immediately off a shoreland.

Lores – The space between the eye and bill in a bird or the corresponding region in a reptile or fish.

Lotic – Pertaining to or living in flowing water.

Macroinvertebrates – Organisms without backbones, which are visible to the eye without the aid of a microscope.

Monocot – Flowering plants, such as grasses, lilies, and palms, having a single cotyledon in the seed, and usually a combination of other characteristics, typically leaves with parallel veins, a lack of secondary growth, and flower parts in multiples of three.

Periphyton – A composite mix of algae, cyanobacteria, heterotrophic microbes, and detritus that is attached to submerged surfaces in most aquatic ecosystems. It serves as an important food source for invertebrates, tadpoles, and some fish.

Permineralization – A kind of fossilization encompassing deposits of minerals within the cells of creatures.

pH – A number between 0 and 14 that indicates if a chemical is an acid or a base.

Plastron – The nearly flat part of the belly or ventral surface of the shell.

Profundal Zone – the area where the bottom sediments are and with no light penetration.

Riparian – Connecting to or located on the bank of a watercourse such as a river or a lake

Rhizomes – A thick plant stem that grows underground and has shoots and roots growing from it.

Scutes – The turtle's shell is covered in scutes that are made of keratin. Individualscutes have specific names and are usually consistent throughout the various species of turtles.

Seep – A place (a type of spring) where water oozes slowly out of the ground.

Submerged aquatic vegetation - A term used to describe rooted, vascular plants that grow completely underwater.

Submerged plants – Same as above.

Succession – the gradual and orderly process of change in an ecosystem brought about by the progressive replacement of one community by another until a stable climax is established.

Terete – (Botany) (esp of plant parts) smooth and usually cylindrical and tapering.

Topography – Graphic representation of the surface features of a place or region on a map, indicating their relative positions and elevations.

Total Suspended Solids – (TSS) It is often listed as a conventional pollutant. TSS are solids in water that can be trapped by a filter.

Transpiration – Transpiration is the process by which moisture is carried through plants from roots to small pores on the underside of leaves, where it changes to vapor and is released to the atmosphere. Transpiration is essentially evaporation of water from plant leaves.

Trophic Levels – The trophic level of an organism is the position it occupies in a food chain.

Turbidity – The measure of the relative clarity of liquid. It is an optical characteristic of water and is an expression of the amount of light that is scattered by material in the water when light is shined through the water sample.

Water Cycle – The continuous movement of water on, above and below the surface of the Earth.

Watershed – The area drained by a river, stream, etc.; drainage area.

Wetland – Land that is saturated with water, either permanently or seasonally, such that it takes on the features of a distinctive ecosystem.

Selected References

Alejandro Villalobos Figueroa and Horton H. Hobbs, Jr. Three New Crustaceans from La Media Luna, San Luis Potosi, Mexico, SMITHSONIAN CONTRIBUTION TO ZOOLOGY • NUMBER 174, SMITHSONIAN INSTITUTION PRESS, City of Washington, 1974

Beletsky, Les, Tropical Mexico: The Ecotravellers' Wildlife Guide, Ecotravellers Wildlife Guides, Academic Press; 1 edition, April 27, 1999

Birds and fish as bioindicators of tourist disturbance in springs in semi–arid regions in Mexico: a basis for management. *Animal Biodiversity and Conservation*, 30.1: 29–41.

Britta Planer-Friedrich, Hydrogeological and hydrochemical investigations in the Rio Verde Basin, Mexico, Institute of Geology, University of mining and technology Freiberg, Technische Universitat, Freiberg On-line Geoscience Vol. 3

Contreras-Arquieta, A. & S. Contreras-Balderas. 2000. Description, Biology, and Ecological Impact of the Screw Snail, *Thiara tuberculata* (Müller, 1774) (Gastropoda: Thiaridae) in México. Chapter 10: 151-160. **In:** Nonindigenous Freshwater Organisms: Vectors, Biology, and Impacts. R. Claudi & J. H. Leach (Eds.). Lewis Publishers, Boca Raton, Florida – London, 2000

Cowardin, Lewis M., et al., Classification of wetlands and deep water habitats of the United States, U.S. Department of the Interior, Fish and Wildlife Service, Office of Biological Services, Washington, D.C. 20240, 1979

Edwards, Ernest Preston, A Field Guide to the Birds of Mexico and Adjacent Areas, Third Edition, University of Texas Press, Austin, TX, 1998

Espinosa–Pérez, H., Gaspar–Dillanes, M. T. & Fuentes–Mata, P., 1993. *Listados faunísticos de*

Grabarkiewicz, Jeffrey D., and Davis, Wayne, An Introduction to Freshwater Mussels as Biological Indicators, EPA-260-R-08-015 November 2008

Hobbs, H. H. Jr., An Illustrated Checklist of the American Crayfishes (Decapoda: Astacidae, Cambaridae, and Parastacidae). Smithsonian Institution Press, 1989

Hobbs, Horton H., Jr. The Subgenera of the Crayfish Genus *Procambarus* (Decapoda: Astacidae). Smithsonian Contributions to Zoology, 117: 22 pages, figures 1-20, 1972

Hubbs, Carl L and Miller, Robert Rush, Six distinctive Cyprinid fish species referred to *Dionda* inhabiting segments of the Tampico Embayment drainage of Mexico, SAN DIEGO SOC. NAT. HIST. TRANS. 18(17):267-336, 2 SEPTEMBER 1977

Impact of leaf-cutting ants on vegetation development at Barro Colorado Island. In Tropical ecological systems: trends in terrestrial and aquatic research GolleyF.G, MedinaE, pp. 99–111. Eds. New York, NY: Springer,1975

Kaufman, Ken, A Field Guide to Advanced Birding, Houghton Mifflin Company, NY, pp. 34-38, 1999

Lemos-Espinal, Julio A. and James R. Dixon. Amphibians and Reptiles of San Luis Potosi. Eagle Mountain Publishing, Eagle Mountain, Utah. 300 pp. 2013.

MARIA GOMEZ-BERNING1, 3, FRANK KÖHLER2 & MATTHIAS GLAUBRECHT1 *Zootaxa* 3381 © 2012 GOMEZ-BERNING *ET AL.* Magnolia Press *Zootaxa* 3381: 1–44 (2012)

México. III. Los peces dulceacuícolas mexicanos. Departamento de Zoología, Instituto de Biología, UNAM, México. http://biblio68.ibiologia.unam.mx/FullText/lf3.html

Miller, Robert Rush Moll, Don, Edward O. Moll, The Ecology, Exploitation, and Conservation of River Turtles, Oxford University Press, 198 Madison Avenue, New York, New York, p. 96, 2004

Official Journal of the State of San Luis Potosi, ANO LXXXVI, Special Edition, Declaration of Protected Natural Area Under the Category of "State Park" Called Media Luna Natural Springs, Secretariat of Ecology and Environmental Management, June 7, 2003

Palacio–Núñez, J., Verdú, J. R., Galante, E., Jiménez–García, D. & Olmos–Oropeza, G., Birds and fish as bioindicators of tourist disturbance in springs in semi–arid regions in Mexico: a basis for management. Animal Biodiversity and Conservation, 30.1: 29–41, 2007

Palacio-Núñez, Jorge, José R. Verdú, Catherine Numa, Daniel Jiménez-García, Genaro Olmos Oropeza & Eduardo Galante, Freshwater fish's spatial patterns in isolated water springs in North-eastern Mexico, Rev. biol. trop vol.58 no.1 San José mar. 2010

Peterson, Roger Tory, A Field Guide to Mexican Birds: Mexico, Guatemala, Belize, El Salvador (Peterson Field Guides), Houghton Mifflin Harcourt, Company, Boston, March 1, 1999

Saha, Amartya K., et al., Effect of leaf-cutting ant nests on plant growth in an oligotrophic Amazon rain forest, Journal of Tropical Ecology, 28:263-270, 2012

Swearingen, J. and K. Saltonstall. *Phragmites* Field Guide: Distinguishing Native and Exotic Forms of Common Reed (*Phragmites australis*) in the United States. Plant Conservation Alliance, Weeds Gone Wild, 2010, http://www.nps.gov/plants/alien/pubs/index.htm

Thomas Popma and Michael Masser, Tilapia - Life History and Biology, SRAC Publication No. 283, March 1999

Villalobos A. and Horton Hobbs Jr., Three New Crustacean from La Media Luna, SLP. Washington, D.C: Smithsonian Institution Press, 1974

W. L. Minckley, Steven Mark Norris, Freshwater Fishes of Mexico, University Of Chicago Press; 1 edition, March 15, 2006

Williams, Jack E., Bowman, David B., Brooks, James E., Echelle, Anthony A., Edwards, Robert J., Hendrickson, Dean A. and Landye, Jerry J. 1985.

Williams, Jack E., David B. Bowman, James E. Brooks, Anthony A. Echelle, Robert J. Edwards, Dean A. Hendrickson and Jerry J. Landye, Endangered Aquatic Ecosystems in North American Deserts with a List of Vanishing Fishes of the Region, *Journal of the Arizona-Nevada Academy of Science*, Vol. 20, No. 1, pp. 1-61, 1985

Websites:

AQUAPLANT, Department of Wildlife & Fisheries Sciences, Texas A&M AgriLife Extension Service, http://aquaplant.tamu.edu

Avibase – The world bird database, Avibase.bsc-eoc.org

Checklist of the Birds of San Luis Potosi, Aves de San Luis Potosi, http://www.birdlist.org/nam/mexico/san_luis_potosi/san_luis_potosi.htm

Festa-Bianchet, M. 2008. *Ovis canadensis*. The IUCN Red List of Threatened Species. Version 2014.3. <www.iucnredlist.org>. Downloaded on 02 March 2015

Hubbs, Carl L - Miller, Robert Rush, Six distinctive cyprinid fish species referred to Dionda inhabiting segments of the Tampico Embayment drainage of Mexico, Transactions of the San Diego Society of Natural History 18: 259-266 (1977)

International Union for Conservation of Nature (IUCN). http://cms.iucn.org/about/work/programmes/species/red_list/search_iucn_red_list/index.cfm

Migration of Birds, Routes of Migration, U.S. Department of the Interior, U.S. Geological Survey, URL: http://www.npwrc.usgs.gov/resource/birds/migratio/routes.htm

Nuisance Muscovy Ducks, Florida Fish and Wildlife Conservation Commission, http://myfwc.com/wildlifehabitats/managed/waterfowl/nuisance-waterfowl/nuisance-Muscovies/

Van Dijk, P.P., Hammerson, G., Vazquez Diaz, J., Quintero Diaz, G.E., Santos, G. & Flores-Villela, O. 2007. *Kinosternon integrum*. The IUCN Red List of Threatened Species 2007: T63671A12705506. http://dx.doi.org/10.2305/IUCN.UK.2007.RLTS.T63671A12705506.en. Downloaded on 19 March 2016.

Printed in the United States
By Bookmasters